"The many methods employed to create new knowledge in the many fields of science and mathematics cannot be described by a simple set of procedures. Over the years, numerous curriculum attempts to do so result more in obfuscation than clarification of the complex nature of creating new understandings, and they do little to assist students in pursuing their own inquiries. Charles Ault has presented this argument with numerous examples in a way that will fascinate the reader."

Joseph D. Novak
Professor Emeritus
Cornell University

"Lumping all disciplines under the single banner 'school science' imperils contemporary science education. To the questions 'What science should be taught?' and 'How is science done?' this book refreshingly and recursively poses a better question: 'Which science?' Ault promotes an appreciation for the rich diversity of scientific knowledge, techniques, and perspectives. This makes for a valuable text for aspiring and practicing science teachers—and those who work with them."

John Settlage
Professor of Science Teacher Education
University of Connecticut

"In an era when standards have become the preeminent metric for assessing educational quality, Charles Ault's *Challenging Science Standards* demonstrates a level of intellectual courage unfortunately rare in academic circles. He argues that the quest for educational standards mirrors the bureaucratic quest for order, measurement, and accountability, a pursuit that with forests destroys biological diversity and reduces a woodland's bounty to only a few products. He suggests that the same thing happens in schools when a false unity is imposed upon the methodological and conceptual diversity of the scientific disciplines. Rather than seeking curricular coherence through the invention of a collection of fictional scientific universals, Ault argues that coherence can be better achieved by inviting students to enact the complexity of the different disciplines in the places where they live. The unity sought by curriculum designers can then be achieved through story, lived experience, and real-life contexts. This is a recipe for enlivening interest, value, and meaning—a recipe that seems more likely to engender powerful learning than complex grids of behavioral objectives."

Gregory A. Smith
Professor of Education
Lewis & Clark College Graduate School of Education & Counseling

"At last we have an engaging, accessible book that challenges how reforms in education that seek unity and standardization within their fields and across all learners have been inconsistent with what we know about children's learning and natural investigations of the world, as well as with the diversity of methods and structures within the sciences. While Ault's engaging stories from his own teaching and research and from his students' observations and experience focus on science, educators of all disciplines will find this a fascinating journey. He honors what children can teach us about the methods of science and the importance of bridging disciplinary structures with personal experience and a sense of wonder about the world."

Carol S. Witherell
Professor Emerita
Lewis & Clark College Graduate School of Education & Counseling

"Beware, conformists. Scientist, historian and author Ault will make you think, and think hard about what conformity does to the motivation of learning. As a fellow educator, parent and student who had the honor of working for and alongside Charles Ault, I find his insights and demand for intelligent education for our children leaving me both breathless and cheering. Management demands may drive the design of education, but Challenging Science Standards: A Skeptical Quest for Unity will surely cause the reader to examine the notion that standardization does not best serve the interests of our children, our families, schools and society."

Thea Weiss Hayes
Teacher & Educational Specialist

"In this important and timely book, Charles Ault leads us on a journey that thoughtfully questions the current focus on supposed 'commonalities' across the different science disciplines. His work reveals the lamentable loss of both the rich context and the unique content particular to individual science disciplines when a unified (and simplified) approach to science is adopted. Numerous classroom and field-based examples (e.g., Pillbug Project) presented in the book highlight the significance of disciplinary science and will prove useful to educators and teacher educators as they embark on a "promising pathway towards understanding scientific expertise in valued contexts."

Nancy G. Nagel
Professor Emerita
Lewis & Clark College Graduate School of Education & Counseling

Challenging Science Standards

A Skeptical Critique of the Quest for Unity

Charles R. Ault, Jr.

ROWMAN & LITTLEFIELD
Lanham • Boulder • New York • London

Pseudonyms appear in place of teacher and school names throughout the book. Classroom vignettes sometimes combine events from different times and places.

Cover Illustration: Specific humidity in a simulation of Hurricane Katrina making landfall made with VAPOR software developed by the University Corporation for Atmospheric Research (UCAR) in Boulder, Colorado. Simulating complex atmospheric phenomena challenges both twenty-first century science and science teaching.

Published by Rowman & Littlefield
A wholly owned subsidiary of The Rowman & Littlefield Publishing Group, Inc.
4501 Forbes Boulevard, Suite 200, Lanham, Maryland 20706
www.rowman.com

Unit A, Whitacre Mews, 26-34 Stannary Street, London SE11 4AB

British Library Cataloguing in Publication Information Available

Library of Congress Cataloging-in-Publication Data

Ault, Charles R., Jr., 1950-
Challenging science standards : a skeptical critique of the quest for unity / Charles R. Ault, Jr.
pages cm
ISBN 978-1-4758-1847-5 (hardcover) — ISBN 978-1-4758-1848-2 (pbk.) — ISBN 978-1-4758-1849-9 (e-book) 1. Classification of sciences. 2. Science—Methodology. I. Title.
Q177.A95 2015
501—dc23

2015020164

∞™ The paper used in this publication meets the minimum requirements of American National Standard for Information Sciences—Permanence of Paper for Printed Library Materials, ANSI/NISO Z39.48-1992.

Printed in the United States of America

To Verne N. Rockcastle
Thanks, Rocky, for being the best science teacher ever.

Contents

Foreword

When people talk about the history of education, one of the most common metaphors invoked is that of the pendulum swing, the idea that schools experience certain trends or fads that arise in one period, disappear, and, years later, surface again in nearly identical form with only a change in terms or slogans. A focus on student needs or personal interests, for example, was prevalent in the progressive educational reforms of the 1920s in the United States, and they appeared again in the student-centered reforms of the 1970s that emerged in reaction to the highly structured disciplinary teaching of the post–World War II period.

While the idea of the pendulum swing, or cyclical thinking, in the history of education can be revealing in some ways, it tends at the same time to obscure a deeper understanding of historical change. Herb Kliebard, the well-known historian of education, offered one of the most thoughtful critiques of the metaphor. He argued that we have more to learn from examining the highly contextual forces and factors that give rise to change in our ideas about schools, teaching, and learning from one period to the next than from some notion of a perennial cycle as a means of understanding change.

In other words, we would do well to explore each period of history on its own terms, looking at the various social and political forces of the moment that, taken together, shift our ideas of what matters.

While the pendulum swing from one educational fetish to another and back again may be more a product of our imagination than grounded in reality, we shouldn't discount the fact that there are nonetheless key trends and phases in the history of education worth paying close attention to. Charles "Kip" Ault's *Challenging Science Standards* is a book that highlights the borders between these important educational trends in a number of interesting ways.

For one, it marks the very beginning of a shift away from universalism and standardization in science education in the United States. Indeed, this is the main focus of the book, showing us with rich examples and lively prose the inadequacies of the current march toward standards.

Professor Ault demonstrates with deft historical awareness how we got to this point and how our current conceptualizations of science have resulted from the need to manufacture representations of science that conform to the demands of the accountability movement in this country. In presenting the case against standards, this book represents, if we give in to the pendulum metaphor (at least superficially), the high point of the swing, the point of pause where society has collectively begun to reconsider whether the current path makes sense—whether there should be a move away from the rush to standardization and its consequences for how science is taught in schools. *Challenging Science Standards* exists at that boundary between educational visions and is a marker, if not a force, in changing our vision.

At a deeper, more contextual level, *Challenging Science Standards* provides a window into the complexity of change. Here the work makes visible the inevitable conflict between two distinct educational trends—one in school organization and governance and the other in the field of science and technology studies—that have unfolded and interacted in ways that fall well outside the pendulum metaphor. Ault's book in this way honors the view advanced by Kliebard that history is never simple and, properly studied, it reminds us of the importance of context at each particular point in time.

This book, although not a work of history per se, owes its existence to the running together of these two particular trends or historical movements. Both were born in the realization of World War II that science had unparalleled value for the nations of the twentieth century. The technological successes of the war (which were visible in things such as radar and atomic weapons) led to the recognition that scientific research and development needed to be nurtured by our country and others across the globe.

The goal since that war has been to somehow ensure that we bend science to national needs (whether they be military or economic). This led, turn by turn, to the increasing focus on educational accountability and standardization we are living with today—to the current focus on standards in science education.

The recognition of the importance of science to the social and political fabric of modern nations at the same time sparked new forms of intellectual studies—the history, philosophy, and sociology of science (together called science and technology studies)—that began to examine the very nature of the sciences as a path to new knowledge. What they found was that, despite the early twentieth-century (and mid-century) desires to cast the sciences as a universal entity, science was instead clearly multifaceted, a collection of different ways of making knowledge rather than some monolith.

The more we have learned about what science consists of—the many diverse intellectual practices that fall under the label "science"—the more untenable the project of standardization has become. *Challenging Science Standards* shows us clearly the collision course these two trends are on and demands that we seek to chart a new course. In this it gives us all hope for a new movement in science education, one more faithful to our understanding of science and how it works.

— John L. Rudolph
Departments of Curriculum and Instruction, History of Science, and Educational Policy Studies, University of Wisconsin–Madison

Author of *Scientists in the Classroom: The Cold War Reconstruction of American Science Education* (New York: Palgrave, 2002)

Acknowledgments

When I completed my teaching practicum, my mentor, Ellen Falk, presented me with a small bowl engraved with the word "Teacher." I have always held that title in high honor. Teaching is both an intellectual and moral calling, and the relationship between teacher and student is always reciprocal. Each grows in the gifts received from the other. I am grateful for these gifts and am indebted to so many teachers and students for the experiences that I drew upon in authoring *Challenging Science Standards*.

Marty and Tom Clark took a chance on me to help them create an innovative school in Colorado in the 1970s. The book reflects their idealism and commitment to independent thinking. I began work at their school as a history major hired to teach in the primary grades, and left as the school's science person. Thank you, Marty and Tom, for the foundation and opportunity that built my career.

My privilege as a science educator has been direct experience in teaching students from age five to fifty-five—from my first years of teaching in the primary grades, through middle-school science instruction, a few years of introductory college earth-science teaching, a stint in elementary-science instruction at a major university, to, finally, teaching graduate students in education at a liberal-arts college. My debt to all of these students—and to the joy they brought to my life—is incalculable. As any teacher knows, students' names are too many to recall but all of their faces are indelibly marked in memory.

Cornell University forged my understanding of learning in science and converted me from an elementary teacher into a science educator. Guided by the paleoecologist John Cisne, I reveled in the astonishing details of nature unveiled by the fossil record of life. D. Bob Gowin introduced me to the discipline of philosophical inquiry, and Joe Novak took me under his wing to sharpen my grasp of the psychology of meaningful learning. My principle mentor, Verne N. Rockcastle, revealed how to find fascination everywhere and turn it to educational advantage. "Rocky" taught me to be doggedly persistent in the connecting of concepts to experiences.

For a quarter of a century, Lewis & Clark College in Portland, Oregon, entrusted me with the coordination of the graduate program in science teaching. My colleagues Carol Witherell, Glenellen Pace, and Nancy Nagel helped me to understand early childhood development, literacy, and integrated education. Their wisdom became my message in my course for secondary teachers: "You never outgrow the need for what is truly elementary," such as a sense of playfulness with ideas. I value the many conversations (and debates) we shared about disciplinary learning, Jean Piaget, and John Dewey.

The idea for this book grew out of several conference presentations and journal publications. John Settlage's support through *Science Education at the Crossroads* was instrumental in sharpening my argument about the problematical persistence of the process approach from decade to decade. John is brilliant both in an academic sense and as a buddy to meet at the pub after a day of intellectual jousting.

The skeleton for *Challenging Science Standards* emerged from Cathy Manduca's invitation to speak in 2012 before an InTeGrate audience ("Interdisciplinary Teaching of Geoscience for a Sustainable Future") on learning to teach the methods of the geosciences. Her endorsement of my work gave me the confidence to speak more loudly and shape the use of geoscience as a case in point about diversity among the sciences. Similarly, Jeff Dodick, formerly of the Hebrew University of Jerusalem, encouraged my analyses of geologic thought as a counterpoint to promoting the habits of mind presumably common to all of the sciences. Jeff put me on the trail of historians and philosophers of science who respect its plurality. I am indebted to Obed Norman of Morgan State University in Baltimore for guiding me through the challenges of making science-teaching succeed with diverse, adolescent populations from economically distressed communities.

As the book took form, feedback from Jerry Kuykendall and Greg Smith, Lewis & Clark Teacher Education faculty, improved my discussions of assessment in science education and the value of a sense of place in civic education, respectively. I am grateful to them both.

Several teachers read early drafts with an eye toward the balance between abstract argument and tangible examples. Thoughts from elementary- and middle-school teacher Thea Weiss Hayes, high-school biology teacher Kate Fisher, and high-school earth-science teacher (and biology, too) Chris Hedeen contributed immensely to this aim. Clearly, the book finds fault with state-level policy and I deeply respect the helpful and responsive reactions of Tom Tinkler and Rob Larson, two very intellectually engaged policy people. I have enjoyed contrasting our perspectives while paddling together with both of them on the Willamette River.

Finding a voice both editorial and storytelling in style was difficult and I thank Sarah Jubar, Rowman & Littlefield editor, in this regard for her excellent guidance. She kept me searching for just the right mix of real-world examples and persuasive rhetoric. Professional input from my other editors at Rowman & Littlefield, Christine Fahey and Lauren Brancato, kept the project on track. I am thankful to them all for supporting me in taking on a somewhat risky task: challenging science standards that enjoy such widespread support.

Science and education permeate the lives of my family members. Family encouragement, as many authors well appreciate, has been indispensable. My son, Logan, reminded me to take breaks and depend upon cycling and paddling to sort through difficult thoughts. Along with the professional editors, I must acknowledge the role played by my daughter, Kori Ault, who read the manuscript as a scientist from outside the field of education. Her suggestions, if I have followed them successfully, have widened the appeal of the book.

I am indebted to my son, Toby Ault, who has established the Emergent Climate Risk Laboratory at Cornell University, for his efforts to help me understand the relationship between climate modeling and the paleontological record of climate variation. The science he is doing is the science the "next generation" will need to grasp.

Finally, to my wife, Phyllis Campbell Ault, I express my heartfelt appreciation for her wise counsel, close readings of each chapter, and the perfectionist touch she never failed to bring to this effort. She has been my companion in a lifelong commitment to teaching and the educational researcher I trust above all others.

Introduction

Halting the Quest

For several decades educators have struggled to identify the attributes all sciences have in common and use these common attributes to guide school science. In the popular mind, this effort constitutes the importance of teaching "the" scientific method. In the policy maker's world, this pursuit yields standards for all Americans that unify the sciences. For teachers, the quest for unity has typically meant teaching science as process.

However, a curriculum that prioritizes what all sciences presumably have in common obscures their vital differences. Studying landslides is very different from doing x-ray diffraction; climate science is unlike medical research. Methods of inquiry in each context correspond to the demands of the problem, and the events of interest vary in fundamental ways: each landslide is unique; all x-rays are the same. Experimenting with humans entails ethical considerations. Climate varies on many scales from decadal to millennial; understanding these variations requires matching virtual worlds to paleontological data.

In brief, the methods for studying diverse phenomena must match their distinctive challenges. That's how the science gets done; that's what makes the sciences interesting and important. To trust—or distrust—the sciences is the key concern. Their portrayal as a common enterprise with universal attributes, as in repeated cycles of educational reform, defeats this aim.

THE QUEST FOR UNITY

Generation after generation, national reforms in science teaching have overemphasized abstract, unifying concepts descended with modification from scientific method. Believing that standardization is prerequisite to educational accountability, reformers promote one-size-fits-all thinking that tends to stereotype science as experimental method. Simply acknowledging that investigators "use a variety of methods," as does the "Nature of Science" section of the Next Generation Science Standards (NGSS),[1] fails to temper the stereotype. Generic depictions of science continue to overshadow *why* investigators use different methods in different circumstances.

Of course, accountability relishes common standards. Without them, comparisons are troubling. Moreover, educational reform tends to equate achievement that meets standards with instructional quality. Achieving standards means progress—unless, of course, the standards are fundamentally flawed. The attempt to frame science standards with a small set of universal attributes is just such a flaw and deserves challenging.

Whether cast from one decade to the next as "the" habits of mind, practices of science, nature of science, scientific processes, crosscutting concepts, or inquiry science, the struggle to unify the sciences has faced an insurmountable obstacle: diversity reigns. Hence, the quest for unity must stumble. Scientific methods are multiple, intelligent adaptations to different kinds of problems. The sciences comprise a loosely federated community of distinct enterprises producing a host of different products. Methodological plurality rules; abstracted unity misleads. "The" scientific method is more myth than teachable skill.

Clearly, scientists use diverse methods. More importantly, scientific methods *adapt* to very diverse challenges: genetics and microbiology investigate the miniscule elements of life; cosmology and geology address physical phenomena on vast scales. Stellar spectra broadcast star temperatures throughout the cosmos; genes maneuvered to arbitrary locations in a fly genome grow legs on heads. Tsunamis reaching Japanese shores confirm great subduction zone earthquakes in Oregon; retroviruses hijack cellular machinery. These discoveries represent the diversity of the sciences, irreducible to a single methodology. Perhaps the geosciences—the oft-neglected orphan of secondary-school science—illustrate this point best.

Democratizing intellect as the habits of mind common to all the sciences has become educational apple pie, the scientific method its simplest recipe. Commitment to the proposition that all sciences are created equal, that they are endowed by their researchers with the same universal properties, that among these are observation and inference, classification, and the control of variables in an experimental design, will not suffice. Accepting the diversity of concepts and methods among the sciences seriously warrants challenging top-level unifiers of the science standards both historically and currently.

Aligning twenty-first-century science curricula with "Next Gen" standards inevitably provokes a sense of déjà vu. Many educators no doubt recall efforts spanning decades to adopt science process skills, follow science's unifying themes and habits of mind, focus on understanding the nature of science, appeal repeatedly to science's big ideas, and teach science as continuous inquiry. Many of these venerable, abstract companions—the skeletal fossils from epochs of unification past—sit at the Next Gen table. They simply have outfitted themselves with new jargon. Under the banners of "crosscutting concepts" and "scientific practices," reformers have reignited the decades-old quest for unity.

THE HOLY GRAIL OF POWER

At the end of the quest for unity lies the holy grail of power over science teaching and learning. This power holds appeal to four constituencies: state bureaucrats, disciplinary scientists, corporate entities, and science educators. Unity advances the self-interest of each one. The four constituencies have found common cause in a conception of the nature of science that distills diverse enterprises into brief, generic abstractions. Allegiance to these lists virtually defines the culture of science education.

State Bureaucrats

States have a compelling interest in the measurability of educational outcomes. The institutionalization of standards, by standardizing expectations, renders complex social and cognitive phenomena "legible," and legibility is prerequisite to measurability.[2] The illegible is unaccountable; state bureaucrats abhor the illegible and most citizens share this fear. They do not trust educational systems to serve all children equitably.

Legibility serves the demand for accountability. Accountability keeps on "biggering and biggering" because, to borrow a phrase from Seuss's Lorax, "everyone, everyone needs" accountability.[3] Given the reluctance to specify science content for all (as well as an inability to agree on what this content would be) and script universal courses of study, standards strive to achieve legibility on a more general plane.

The bureaucrat's ideal curriculum standardizes the nature of science (or the processes of science), independent of context. Legibility trumps diversity; state interests displace personal ones. Test scores signify learning, and policy unfolds based upon interpretations of these scores. But the tests are not learning; they are legible indicators of learning:

> Unfortunately, making things legible inevitably entails making them simpler. Ultimately, the legible representation begins to stand for the real thing as abstractions replace realities, and the arranged order of the virtual world supplants the untamed complexities of the organic.[4]

Disciplinary Scientists

In the nineteenth century biologists ("naturalists" in the language of the era) earned a seat at the table of science and claimed equal stature with chemists and physicists. They argued that their science was as scientific (and as experimental) as any other. For a while, with the discovery of deep time, geologists joined them. For most of the twentieth century, especially in terms of school science and general science requirements at prestigious universities, the earth sciences (and their space science cousins) received their invitations to join the party on a space-available-only basis.

Recent critics of climate science tend to disparage modeling as credible science. Modelers run experiments in virtual worlds. Planetary scale geo-experiments, at least intentional ones, are taboo (incidental oil spills are not). As Kim Kastens and Cathryn Manduca recently wrote, "As a discipline the geosciences often struggle to find a place at the scientific table."[5]

Philosopher of science (especially physics) Nancy Cartwright has pictured the sciences as more of a patchwork than a pyramid. For Cartwright, disciplines select tools and seek imagery respectful and responsive to "life in the messy world that we inevitably inhabit."[6] She warns against choosing the "wrong tools with which to change it."[7] Reducing disciplinary diversity to a set of universals raises the risk of using the wrong tools.

Some even argue that a pervasive disunity prevails among the sciences. Based on a comparison of the work done by particle physicists with that of molecular biologists, especially in terms of laboratory culture, sociologist Karin Knorr Cetina concluded that the sciences comprise not a single enterprise, but instead "a whole landscape—or market—of independent epistemic [knowledge-producing] monopolies producing vastly different products."[8]

The characterization of science as universal facilitates the argument that "my science is as good as your science." Disciplines are deemed equally scientific when demonstrated to conform to the same criteria. For example, zoo-archaeologists attempting to gain entry to public schooling use this reasoning to justify having middle schoolers investigate animal remains found in local middens in order to learn how people have inhabited the land through time (see chapter 6, Archaeological Outlier section).

They certainly may hold equal commitments to rationality and empirical truth. At the same time this commitment to rationality requires that methods of inquiry diverge to solve diverse problems. Disciplines earn stature by the value of their accomplishments, not by exemplifying a universal idea of scientific method. Unfortunately, the quest for unity satisfies the thirst for stature.

Belief in the unity of scientific enterprises justifies assessment that substitutes generic, one-size-fits-all characterizations of science for messy, diverse disciplines where thinking adapts to context in distinctive ways. By oversimplifying the structure of disciplines and the characteristics of thinking adapted to particular problems, the universalist agenda and the bureaucratic drive for legibility misrepresent authentic inquiry.

Corporate Entities

The institutionalization of national educational standards—their dissemination and implementation—promises lucrative returns to publishing, training, consulting, and testing industries. Achieve, Inc. (a nonprofit) currently leads these efforts. Staff and curriculum development are costly.

Spending and earning are not bad things, of course, especially if students prosper as a result. Nevertheless, because measuring makes money, special interests accumulate in the wake of reforms, giving them a momentum that makes skepticism increasingly difficult to market. A host of enterprises are creating materials for aligning classroom teaching with new standards. Institutions seeking funding for projects to advance science education have no choice but to cast their proposals in terms of the NGSS. For-profit and nonprofit providers of professional development, school district trainers, and consulting firms wait in the wings eager to help.

Science Educators

Philosophers of science, struck by the stupendous achievements of experimental physics and chemistry over multiple centuries, have attempted to discern the reasons for such progress. Seeking nothing less than the distillation and bottling of the elixir of scientific practice, they examined deductive and inductive reasoning, the logic of hypothesis testing, and the mathematization of theory.[9]

Science educators ran with this tradition. It gave common purpose to schooling. Simplified, the philosophy of universal science undergirded the promise of democratizing intellectual rigor—of transforming minds with habits of thought credited with the progress of civilization. Jean Piaget predicated his stages of growth in logical reasoning on exactly this progress. Many classroom teachers have joined the ranks of science educators enamored with Piagetian reasoning and decontextualized inquiry skills.

Science educators believe in the responsibility of K–12 schooling to prepare students for lives of social responsibility. They must respond to the pressure to keep the pipeline of future engineers and scientists full and open to all, generation after generation. A simplified agenda serving both causes is guaranteed to receive immense attraction. Catalyzed by Dewey's 1910 essay, "Science as Subject-Matter and Method" (examined in chapter 2), the science education community has pursued the holy grail of unity with fervor. Even though each cycle of reform reveals its shortcomings, the quest continues.

As philosophers, historians, and sociologists of science dig deeper into the practices of the sciences, dispelling the myth of "the" scientific method, science educators have purported to pay attention. However, generic, simplified lists of universals are deeply ensconced in the profession. In effect, science educators fear that letting them go will encourage teaching to revert to separate subject, rhetoric of conclusion, and right-answer norms that are straightforward to teach and to test.

Science educators, in the creation of curricula and the training of teachers (elementary through secondary), feel called upon as guardians of the quest for unity among the sciences. Their professional identity—an identity setting them apart from other professors of education, for example—depends upon this cultural norm. Ideas about the nature of science and the culture of science now pertain more to the community of science educators than to that of scientists. These ideas equate in many minds with critical thinking, inquiry skill, and the development of intelligence.

All four constituencies have good reason to combine their energies in crafting and institutionalizing science teaching predicated on the value of the eight practices and seven crosscutting concepts of science codified in the NGSS. Implementing this agenda advances the interests and stature of each group.

The prospects for admitting to flaws in the standards and charting quite different directions are daunting. Challenging the standards must be fundamental. In fields as diverse as bioengineering, materials science, climate modeling, systems thinking, and medical research, active citizen engagement in both policy decisions and recruitment of future scientists depends on doing better than in the past.

RESPECT FOR DIVERSITY

The sciences are plural; a singular, scientific method is myth. The challenge to match methods to concepts and adapt both to the demands of problem-solving plays out in many different contexts. Some approaches isolate subsystems and test hypotheses of causal relationships among a discrete number of variables.

Others accept complex phenomena as a given, irreducible to laboratory manipulations. Investigations with human subjects must adhere to ethical standards and seek informed consent. Field studies in the life and earth sciences develop protocols suited to investigating the natural world on its own terms. Rhetorical style, explanatory ideals, and trusted methods vary from one discipline to another. The sciences adapt both conceptually and methodologically to diverse challenges. No one approach serves all purposes.

The portrayal of the sciences as unified under the banner of experimental method accompanied mid-nineteenth-century proposals for educating the next generation of scientists. One science was deemed as good as any other science to the extent it practiced controlled experimentation leading to rigorous predictions. Biologists argued that their methods were as equally scientific as those of the physicists. Biology could not be reduced to mere description. Its "comparative method" responded to the complexity of life, enabling the reconstruction of extinct creatures from fragments of their remains.[10]

According to science historian and paleontologist Stephen Jay Gould, the triumph of Charles Darwin's approach to inferring ancestry and the origin of species, as well as Charles Lyell's demonstration of the mutability of the earth on vast scales in time and space, gave birth to a second style of science: the historical ("experimental" being the first).

The "contingency" of events in time, argued Gould, made precise prediction an unreasonable goal. At the same time, the existence of complex, historical artifacts could be depended upon to trace causal connections.[11] These artifacts, both as the fossil record and accumulated changes in DNA, offer a Darwinian account of how baleen whales came to be from toothed, limbed, and un-fluked proto-whale ancestors (some who walked). This story of descent functions very well as an explanation but remains mostly silent on what whale progeny may become ten million years from now.

The historical style of science often addresses complex, singular objects unfolding through time: one earth, one universe, one history of life. Whether applied in geology, paleontology, evolutionary biology, or even cosmology, the historical style reduces uncertainty and justifies causal claims. It responds to the challenge to solve problems that cannot be addressed by experimental methods.

Often scale presents an obstacle that methods of inquiry must overcome. The vast scale of geologic time is a perfect example. Geologic reasoning depends upon finding evidence of temporal relationships, and scale, both temporal and geographic, is an obstacle. Establishing sequence and synchrony are key elements in geologic arguments. However, geologic clocks are subtle and often absent, and the remains of the past are haphazardly distributed. Dating is difficult, yet essential. Good geologic arguments depend upon reliable temporal data, and scale presents an immense challenge—a challenge that almost defines geoscience.

The case of geologic inquiry and the contrast between historical and experimental styles provide starting points for exploring how disciplines vary. Unless properly ordered in time, geologic (and evolutionary) explanations lack validity. When solutions to problems on different scales cohere, geologic arguments gain credibility. In a parallel fashion, experimental methods rely upon controlling variables to achieve this aim.

Whether implicit or explicit, even the metaphors embedded within disciplinary thought vary. Metaphors and analogies jump-start disciplinary engines whether inquiries play out in the laboratory, the field, or virtual worlds: the "clockwork" universe, the earth as an "organism," the "tree" of life, the "hardware" of the brain. Recognizing the influence of metaphors and analogies reinforces the insight of disciplinary diversity.

RESCUING THE NGSS

Respect for the diversity of scientific enterprises sets a very different anchor for curriculum design than does the quest for unity. Making diversity the starting point fundamentally changes answers to the question, "What to teach?" Respecting diversity can rescue the NGSS's overly ambitious portrait of scientific unity through careful development of core disciplinary ideas and their assessment in terms of disciplined—not universal—approaches to thinking and doing science.

The progression of core ideas ought to culminate in mastery of a small number of conceptions that organize thinking in well-defined contexts, keeping concepts linked to methods of inquiry in each step. Attempts to subsume all lessons in science within the same small set of abstractions ("scientific practices" and "crosscutting concepts") detours progressions from this goal, divorces thinking from doing, and devalues diversity.

For example, thinking about the earth in space profits from a good grasp of the concept of "frame of reference," which encompasses ideas such as "motion as a change in position." Developing a powerful invention of mind such as "frame of reference" may begin with elementary-school talk about "staying where you are." That's a problematical thing for a child to do if sitting in a wagon pulled by another child along the sidewalk. From the frame of reference of the wagon, the child has not moved. From the frame of reference of a place on the sidewalk, the child in the wagon has moved.[12] Analogously, the sun moves across the sky daily from the frame of reference of an earthly observer, but for someone in orbit, what's moving and what's not moving appear quite different.

Frame-of-reference thinking helps to sort through explanations of lunar phases and even the predictions of relativity theory. That's a remarkable progression that spans the disciplines of physics and astronomy. However, there is no need to place rolling wagons, lunar phases, and galactic scale motions in the context of universal, crosscutting concepts. The better approach is to develop curricula that progressively develop conceptions for solving different kinds of problems while at the same time introducing methods of inquiry that correspond to such thinking.

Understanding relative motion is just one example. Generic categories of scientific practices and crosscutting concepts drop out of the picture; learning to use the concept of "frame of reference" to carry out investigations in multiple contexts has sufficient worth without them. Time invested in aligning science teaching with generic abstractions is wasteful effort.

Homology is another example of a conception that has stood the test of time as an organizer of thinking *and* doing in appropriate contexts. Biologists, even well before the publication of *The Origin of Species*, were familiar with the idea of homology—same structures, same relative positions, but differently modified from one species to another. Post-Darwin, biologists have searched for genetic and anatomical homologies in order to infer common ancestry. The classic example of the corresponding bones in the bat, pterosaur, bird, and human forelimb has permeated textbooks for some time.[13]

The conception of homology perfectly illustrates the union of thinking and doing in the life sciences, both historically and at present. It's an idea that guides method and an idea whose meaning has been modified by inquiry. In the modern sense, homologies, whether of limbs or DNA sequences, provide clues to the degree of common ancestry shared by different species.

On a trip to the zoo, comparing the limb structures of storks, sea lions, rhinos, zebras, bears, goats, antelopes, giraffes, penguins, elephants, bats, orangutans, lemurs, and even the African lungfish, has the potential to provoke reflection on evolutionary relationships as well as adaptations suited to specific habitats.[14] When thinking in terms of homology while sketching limbs (a practice used by animators from Disney studios), concept and method row in the same boat.[15]

With diversity properly respected, learning in science may ascend distinct ladders of strategies tailored to meet disciplinary challenges. Planning and assessment correlated with these ascents can liberate science teaching from the persistent myth of the scientific method. There are many methods of inquiry adapted to diverse phenomena—and that's the challenge science standards ought to address.

Challenging Science Standards voices skepticism toward the quest for unity through successive eras of reform in science education. Using analyses of curriculum, assessment, and classroom learning within historical context, the argument honors the diversity of scientific enterprises by recognizing the interdependencies between conceptions and their corresponding methods of inquiry.

Regrettably, the shortcomings of the NGSS are being institutionalized through state-by-state adoptions, perhaps irreversibly. There are solutions to these shortcomings. First, there is a need to acknowledge the limitations of the standards and their mischaracterization of scientific unity. Second, there is the work needed be done in constructing curriculum from a substantially modified perspective, one that simply calls a halt to framing all science learning in terms of universal categories. Then the messy, busy, and no doubt controversial tasks of making value decisions about what to teach may begin in earnest, guided by core progressions that truly honor disciplined thinking and doing.

Respecting disciplinary diversity promises no simple, generic formulas for teaching all sciences. It does offer a starting point on a promising pathway toward understanding scientific expertise in valued contexts.

NOTES

1. NGSS Lead States, *Next Generation Science Standards: For States, by States* (Washington, DC: National Academies Press, 2013). For current information about the NGSS, visit http://www.nextgenscience.org/.
2. James C. Scott, *Seeing like a State* (New Haven, CT: Yale University Press, 1998).
3. Theodore Geisel (Dr. Seuss), *The Lorax* (New York: Random House, 1971).
4. Charles R. Ault, Jr., and Jeff Dodick, "Tracking the Footprints Puzzle: The Problematical Persistence of Science-as-Process in Teaching the Nature and Culture of Science," *Science Education* 94 (2010):1109. This passage interprets one of James Scott's principle theses.
5. Cathryn A. Manduca, and Kim A. Kastens, "Geoscience and Geoscientists: Uniquely Equipped to Study Earth," *Geological Society of America (GSA) Special Papers* 486: 1. doi: 10.1130/2012.2486(01).
6. Nancy Cartwright, *The Dappled World: A Study of the Boundaries of Science* (Cambridge, UK: Cambridge University Press, 1999): 18.
7. Cartwright, 12.
8. K. Knorr Cetina, *Epistemic Cultures: How the Sciences Make Knowledge* (Cambridge, MA: Harvard University Press, 1999): 4.

9. Robert Frodeman, "Geological Reasoning: Geology as an Interpretive and Historical Science," *Geological Society of America (GSA) Bulletin* 107 (1995): 960–68.

10. An interesting window into the foundations of debate about a universal science and its implications for science education can be found in Thomas H. Huxley's 1845 essay, "On the Educational Value of the Natural History Sciences," in T. H. Huxley, *Science and Education* (New York: Collier and Son, 1901).

11. Stephen Jay Gould introduced his notion of the second style of science in a 1986 essay, "Evolution and the Triumph of Homology, or Why History Matters," *American Scientist* 74 (1986): 60–69. By "homology," he meant, in effect, application of the comparative method developed by the French anatomist Baron Georges Cuvier in the early 1800s.

12. The frame-of-reference lesson featuring a child told to "stay where you are" while sitting in a moving wagon appeared in the fourth-grade unit, "Relative Position and Motion" of the *Science Curriculum Improvement Study* (SCIS) developed in the 1960s under the leadership of physicist Robert Karplus (University of California—Berkeley, 1970) and reissued under the direction of Herbert D. Thier as *SCIIS* (Chicago: Rand McNally, 1978). The program adhered to Piagetian notions of the development of intelligence.

13. For decades the Biological Sciences Curriculum Study's (BSCS) high-school textbook, *Biology: An Ecological Approach*, has featured the role of "homology" in classification and inferring evolutionary relationships. See BSCS, *Biology: An Ecological Approach,* 10th Ed. (Dubuque, IA: Kendall Hunt, 2006).

14. One such activity can be found in *Z.E.S.T.—Zoos for Effective Science Teaching*, a project of the Bronx Zoo Education Department funded by the National Science Foundation (NSF Grant No. TPE 8751488, 1989).

15. Inspiration and guidance for limb sketching on a zoo tour mentioned in the "Introduction" and addressed in chapter 14, "The Elbow and Ankle Tour of the Zoo," stems from Z.E.S.T. (previous note) and other sources: Chuck Jones, "How to Make a Tennis Shoe for a Percheron," in *Chuck Amuck: The Life and Times of an Animated Cartoonist* (New York: Farrar, Straus and Giroux, 1989); Stuart Sumida and Beth Rega, "Anatomy and Animations: Combining Classic Techniques with Cutting-edge Technology," NSTA Regional Convention (Portland, OR, 2002); Neil Shubin's *Your Inner Fish: A Journey into the 3.5-Billion-Year History of the Human Body* (New York: Pantheon, 2008).

1

THE NATURE OF THE PROBLEM

> There abounds a false idea of knowledge. This false view leads to a debilitating rigor.
> This obsessive rigor leads to a silencing: questions of fundamental interest are forbidden.
> Therefore, questions not asked result in a false idea of knowledge.
> *The cycle is safe and therefore popular.*
>
> —D. Bob Gowin[1]

Each epoch of reform in science education recycles the steps and language of the scientific method. Even calls to teach the nature and culture of science dwell upon the imagery of science as generic process. Each cycle of reform reinforces the popularity of this imagery. This false view of science, represented as national standards, has silenced fundamental questions about science teaching for all.

NOTE

1. D. B. Gowin, "The Learning Environment, 1986, at Cornell University: Ten Questions in Search of Answers," Unpublished manuscript (Ithaca, NY: Cornell University, 1986). Gowin is best known for his book *Educating* (Ithaca, NY: Cornell, 1981) that elaborates carefully upon the distinction between thinking and doing in the context of inquiry.

1

Unity v. Diversity

Working in tandem, methods must match concepts in adapting to the challenges of inquiry. This match calls for forms of expertise that differ substantially from one field to another. Actual scientific work, tailored to different circumstances and explanatory aims, bears delicious fruits that appeal to different tastes.

The tools of the trades and their astonishingly varied products spark interest. The sciences adapt diverse methods to address diverse phenomena, responding to distinct challenges with methods matched to concepts. The quest to unify the sciences and abstract their common aspects as drivers of school science subverts this fundamental truth.

FALSE UNITY

"The" scientific method and its abstract descendants claim to represent universal, underlying assumptions shared by all the sciences. However, neglecting the fundamental truth of diversity cultivates ignorance while masquerading as insight. Glossing over differences in the quest to unify contributes to the public's misunderstanding of what scientists actually do—and when and why scientists merit trust or distrust. While all sciences may act to reduce uncertainty, no single method works best to do so.

John Rudolph, editor of the prestigious journal *Science Education*, argues that special-interest groups readily exploit the public's misunderstanding of science and how different fields address uncertainty.[1] For the general population, a carefully prescribed experimental method promises truth.

The public mind perceives uncertainty in claims about the climate, both its past and its future, as the result of poorly conducted science. For example, climate-change skeptics tend to disparage climate modeling as a departure from the scientific method. As a result, the public fails to appreciate the methods of climate scientists as appropriate ways to represent complex phenomena.

Failure by educators to couple the methods of investigation to the demands characteristic of the problem—and to bear witness to how thinking adapts accordingly—invites criticism of specific content statements in the science standards. For example, in West Virginia the state Board of Education first altered, and then withdrew, changes to the Next Generation Science Standards (NGSS) statement regarding global temperature. The board wished to have students ask questions about the evidence for the "fall" as well as the rise in global temperatures over the past century.[2]

Simply arguing about evidence, however, comes up short of appreciating how the science gets done. That failure enables critics to focus attention on curriculum reduced to "a rhetoric of conclusions."[3] Objectionable ones draw politicized ire.

Departures from the stereotype yield suspicious findings; therefore, belief in the experimental stereotype makes uncertainty in the work of climate scientists an inviting target. Even those convinced that global warming is under way may misunderstand the nature of climate science. For example, single extreme events (Hurricane Katrina, extreme drought in California, Superstorm Sandy) cannot, in principle, signal the effects of global warming. Their frequency against a backdrop of variation on a millennial time scale does.

Claiming climate change requires collating the frequency, intensity, and distribution of storms and droughts temporally and geographically on multiple scales. Such an approach to inquiry—at best still only part of the story—differs greatly from manipulating a single variable or weighing the relative contributions of several variables in order to determine cause-and-effect relationships. Sampling the past presents climate scientists with unique challenges; their forecasts of the future are necessarily probabilistic.

Sampling and statistical analysis permeate medical research as well as climate science. Carefully controlled experiments (the "gold standard") do occur in medicine, but human variation and ethical standards present special challenges to doing medical experiments. Vulnerable populations (children, for example) are often underrepresented in randomized control trials.

Sometimes inevitable uncertainty translates into reversals of expert advice—for example, with regard to the benefits according to age cohorts of prostate antigen tests for men and mammograms for women. Medical advice reversals create a public relations nightmare. Progress in physics, on the other hand, does not seem to face such stark reversals.

Medical researchers and climate scientists must match methods to concepts in ways that best serve their interests. The failure to teach the sciences from the perspective of plurality—from the point of view that disciplines are distinctly different enterprises, responding appropriately to very different challenges—has the potential to invite doubt and mistrust.

Clearly, reform movements over the ages have espoused teaching science as process, and they elevated concern for understanding the nature of science. The current iteration of reform, expressed in the NGSS (the twenty-first century's replacement for the moribund National Science Education Standards [NSES] of the 1990s),[4] anchors science teaching to the *practices* of science and its *crosscutting concepts* (and to engineering design as well, but that is not the subject of this argument). Admirably, the NGSS emphasize how process skills and subject knowledge intertwine. Such an acknowledgment tempers efforts to teach universal processes of science—observing, predicting, classifying, and controlling variables—in isolation.

In the end, despite some promising attention to how skills and knowledge intertwine, scientific unification, independent of discipline, still triumphs in the NGSS. As a result, two of the three dimensions framing the standards trumpet generic notions of science: a list of eight practices common to all the sciences (questioning, modeling, investigating, interpreting data, using mathematics, explaining, arguing from evidence, and communicating information) set in tune to seven concepts that cut across the disciplines (patterns, cause and effect, scale and proportion, systems and models, energy and matter, structure and function, stability and change).[5]

"Modeling" rates mentioning twice perhaps because it topped the pre-NGSS list of "integrated process skills."[6]

"Core disciplinary ideas" comprise the third dimension of the standards and reflect subject-by-subject objectives for student achievement. The abstract NGSS dimensions of "scientific practices" and "crosscutting concepts" frame the evidence needed to assess student achievement:

> The evidence statements provide detail on what students should know and be able to do in order to satisfy each performance expectation (PE) at the end of instruction. Given that each PE is three-dimensional, the statements describe how students can use the practices, crosscutting concepts, and disciplinary core ideas together to demonstrate proficiency on the PEs by the end of instruction.[7]

Take, for example, the core idea from the earth sciences, "weather and climate." High-school students are expected to be able to describe how climate responds to changes in the flow of energy through earth systems. In charting such expectations, practices occupy the column to the right and crosscutting concepts occupy one to the left of each block of declarative statements (or "rhetoric of conclusions") constituting a core idea such as ESS2.D:

> The foundation for Earth's global climate systems is the electromagnetic radiation from the sun, as well as its reflection, absorption, storage, and redistribution among the atmosphere, ocean, and land systems, and this energy's re-radiation into space.[8]

The practice that frames this core idea is "developing and using models." The crosscutting concept is "cause and effect."[9] Both dimensions are generic—there are no statements about how climate scientists build models or what the challenges are to establishing cause and effect when studying complex earth systems. In other words, the practices and crosscutting concepts that frame performance expectations lack reference to genuine, disciplinary understanding. Core idea ESS2.D remains an inert proposition.

For good measure the NGSS have promulgated eight basic understandings of the "Nature of Science" (NOS): four are associated with the practices, and four pertain to the crosscutting concepts.[10] The NOS understanding pertinent in the energy and climate example is that "science knowledge is based on empirical evidence." This universal principle glosses over the difficult question, "How?" Answers to "How?" and "What counts as evidence?" vary with context and depend upon conceptualization.

Evidence does not stand alone. The mischaracterization of climate science encouraged by framing its core ideas with universal constructs thus debases debate. The lists of eight practices and seven crosscutting concepts—even when aided and abetted by understandings of the nature of science—cannot transform declarative statements into intellectual accomplishments. The core disciplinary ideas break disciplinary knowledge apart into propositional statements. All the NGSS phrases and all the NGSS themes cannot put Humpty Dumpty back together again.

With its disciple, assessment, ensconced in the pulpit, the NGSS adamantly proselytize scientific unity. Achieve, Inc., the nonprofit corporation working with the states to implement both the Common Core and the NGSS, has fashioned a scoring system that awkwardly bends curriculum into alignment with these new standards: Educators Evaluating the Quality of Instructional Products (EQuIP).[11] EQuIP works hand-in-glove with Next Gen Performance Expectations. EQuIP focues on the alignment of lessons and units with the three dimensions of the NGSS. A forthcoming, parallel rubric, Primary Evaluation of Essential Criteria for Alignment (PEEC), centers on the task of aligning year-long science programs and materials with the dimensions of the NGSS. (A draft of the PEEC was released in May 2015 at www.nextgenscience.org/ngss-peec-alignment.)

To be fair, the NGSS feature core disciplinary ideas arranged in reasonable grade-level progressions. For example, the "sophistication of student thinking" about "the universe and its stars" in middle grades examines the idea that "the solar system is part of the Milky Way, which is one of many billions of galaxies."[12] The statement is tangible—and no doubt fodder for test construction. Lunar phases, earth's rotation, orbital motion, stellar spectra and luminosity, celestial objects, and the big bang all receive their due in this progression. Yet they earn a place in the curriculum only if able to keep to the three-step polka beat of core-practice-crosscut.

Achieve, Inc. and the NGSS do not hide their intent to unify. Achieve "in no way endorses separating the disciplinary core ideas from the other two dimensions."[13] Learning, for instance, about stars and galaxies and the evidence of their cosmic past serves a higher calling than grasping the structure of the universe. Mastering the NGSS's scientific worldview matters most. This worldview of scientific unity purports to justify schooling in science for all.

Adopting the NGSS presumes the value of having students adopt the NGSS's characterization of the sciences: *E pluribus unum* (Latin for "One from many")—and a strange *unum* at that, with three separate lists of universals. The NGSS's lists of generic abstractions are meant to dispel the stereotype of the scientific method and the reduction of science to process. Ironically, the short lists of generic abstractions reinforce the quest to define a universal science. In place of "the" method, the NGSS offers perpetual cognitive motion.

Scientific practices and crosscutting concepts power disciplinary core ideas to generate truth in keeping with the nature of science. Unfortunately, this worldview masks the actual work of myriad disciplines often only loosely united to each other.

The NGSS foster a deceptive plurality and a false unity. The NGSS's generic lists oversimplify what scientists do and disturbingly promote ignorance of how diverse methods and distinctive concepts combine to achieve different ends. The standards are not curricula. However, a curriculum must align with NGSS standards as scored by EQuIP and PEEC rubrics. Unless cast in terms of the generic dimensions found at the summit of the NGSS, lesson plans and curricula proposals will receive low scores, whatever innovative ties to actual science it might embody. The evaluation process shoehorns all sciences into a one-size-fits-all framework.

The same scientific practices and crosscutting concepts frame performance expectations for students. The lists do lend themselves to classroom display. No science should be taught divorced from the two dimensions of practices and crosscuts, claims Achieve, Inc. Picture glossy posters listing the eight practices and the seven concepts with an array of sponsoring institutions' logos prominently printed at the bottom—or perhaps just "Achieve, Inc." Names that start with "National," as in the "National Science Teachers Association" and the "National Academy of Sciences," conflict with the NGSS subtitle: *For States, by States*. No one wants to impose a national curriculum; everyone hopes to achieve.

Properly positioned, such posters afford students constant reminders of the highest aims their curriculum (and everyone else's) serves. The dimensions of the NGSS presumably overturn the myth of the scientific method. As the poster children of science in the twenty-first century, two of the three siblings mostly re-create the myth.

DECADES OF MISLEADING FOOTPRINTS

Reinventing the scientific method through a cycle of reform has happened repeatedly. In 2008 *Teaching the Nature of Science through Process Skills*, by well-respected science educator Randy Bell, captivated the reform market with a wealth of well-received inquiry exercises for upper-elementary- and middle-school grades.[14]

A decade previous to Bell's book, the National Research Council published *Teaching about Evolution and the Nature of Science*, popular among high-school teachers.[15] In their textbook integrating science teaching with a good grasp of the principles of multicultural education, John Settlage and Sherry Southerland advocated teaching prospective primary-grade teachers to consider science as a culture typified by a shared commitment to the basic scientific processes of observing, inferring, classifying, measuring, and communicating.[16]

All three works posit a small set of universal abstractions as the ultimate goal for learning science: scientists use diverse methods, scientific conclusions are tentative, and science is a way of knowing. Not surprisingly, all three books share an exercise based on intersecting footprints (representing dinosaurs in the books by Bell and the NRC; animals in the snow in Settlage and Southerland) that first entered the science classroom in the 1960s.[17]

The exercise calls upon students to separate observations from inferences in a stylized rendition of actual fossilized trackways (possibly of a predator dinosaur tracking prey somewhere in what is now modern-day Texas). Developing a strong distinction between observing and inferring has permeated materials for teaching science as process for a great while at all levels of instruction; the footprints puzzle provides a telling example.[18]

Repeated attempts to depict the sciences as unified by method, process, or nature are the source of déjà vu in the NGSS practices and crosscutting concepts. The American Association for the Advancement of Science (AAAS), with guidance by Robert Gagné, introduced the list of processes common to the sciences in the 1960s.[19] NOS settled into a list of understandings in the 1990s. Bell linked the nature and process lists in 2008.

Unity has long lacked unity. The original National Science Education Standards (NSES) listed four unifying categories of standards separate from traditional disciplines: unifying concepts and processes, science as inquiry, history and nature of science, and science in personal and social perspectives.[20] The NGSS collapsed these four domains into two of its three dimensions (plus NOS). Achieving unity never seems to work well, and the struggle comes at a high price. The effort to do so has left tracks for many decades.

DIVERSITY REIGNS

Different obstacles arise when studying different phenomena. Determining chemical reaction mechanisms that take place at minute scales and rapid speeds challenges the chemist. Geologists attempting to explain the structure of river deltas and sediment fans laced by submarine canyons on a continental shelf must coordinate processes spanning immense amounts of time with those that are sudden and catastrophic. They face dramatically different challenges.

Furthermore, the organic chemist's carbon atoms are indistinguishable; the marine geologist's seafloor fans and river deltas are unique. Atoms are tiny; deltas are quite large. The science of carbon chemistry and the science of delta formation adapt to these different characteristics (see table 1.1). These differences in characteristics have

Table 1.1. Comparing Characteristics

	Scale	Identity	History	Prediction	Concept Boundary	Inquiry Style
Carbon Chemistry	Tiny	Uniform	Unknowable	Precise	Rigid	Experimental
Delta Formation	Large	Unique	Essential	Probabilistic	Fuzzy	Interpretive

profound consequences for thinking and doing, from recording data to constructing explanations.

Of course, there are different isotopes and energy states of carbon, and deltas from one continent to another have similarities. However, there is no way to decipher one carbon atom's history from another's. The opposite is true for deltas (or, in more sophisticated terms, seafloor fan-canyon complexes), and this history is essential to explaining how they came to be.

Chemical explanations assume the constancy of particle identity—protons are protons, electrons are electrons, carbon 12 is carbon 12. Geologic explanations zero in on uniqueness—how one delta differs from another and why. Currents differ from place to place, and some places, but not others, are subject to earthquake-induced, underwater landslides.

The boundary between the concepts of carbon and nitrogen is well defined. The boundary separating the concepts of delta from "bar at the mouth of the river" is fuzzy. One example of a carbon atom is as good an example as any other. The features of a continental shelf at the mouth of one river always differ in some respects from the features at another.

As phenomena vary, so do the challenges to inquiry, with consequences for concept formation and investigative methods. In chemistry, ambiguity (any inability to tell elements clearly apart) is anathema. In geology, ambiguity (the ability to draw the line between delta and bar) is the side effect of compare-and-contrast reasoning. It is the inevitable price paid for categorizing phenomena with unique histories.

This way of thinking—how disciplines differ in fundamental ways—can be difficult to grasp because it's not the norm in science education. Teachers are quite comfortable thinking about teaching students to observe, infer, predict, collect, and analyze data. Nearly everyone in science education talks in those terms and has for over a hundred years.

Few teachers talk to each other about how waterfall phenomena are categorically different from infectious diseases or how carbon is unlike a river. The consequence of such conversations would mean at least tempering, if not abandoning, the quest for scientific unity and its stature at the apex of curriculum design.

Each science depends upon methods and ideas adapted to solving its own problems.[21] In medicine, researchers invoke the protocol of a double-blind study with random assignment of patients to treatment or placebo. The explanatory ideal is to determine the efficacy and safety of the treatment. Medical researchers arrange clinical trials in keeping with their knowledge of human variation. Does the treatment stop the cancer without harming the patient? Do the benefits outweigh the risks?

When the patient is the earth and the treatment is the carbon-based global economy, there is only one subject in the study: one earth, one history—but many possible futures. The protocol of double-blinding and the random assignment of participants to meet statistical challenges play no roles. When has the earth cooled or warmed, and how rapidly did it do so during the Holocene Epoch? What future is now likely? There is no large population of earths to draw samples from randomly and apply controlled treatments to, in order to find out.

The Challenges of Climate Science

Climate scientists design investigations adapted to their understanding of climate and the obstacles to its study, not in keeping with the simplistic belief in a generic scientific method. For example, glaciers in Antarctica and Greenland store samples of air that span hundreds of thousands of years. Core samples reveal that the carbon dioxide content varies cyclically with time. Correlating these levels with average global temperatures indicates that air rich in carbon dioxide corresponds to warm periods.

In addition to the bubbles in glacial ice that track the composition of the ancient atmosphere, Earth harbors several ghostly thermometers. For example, properly calibrated, oxygen isotope ratios in the planktonic shells of deep-sea muds, the remains of Pacific coral reefs, and even cavern limestone, encode the earth's temperature history. Sampling them is quite a challenge. These strange thermometers are referred to as "proxies." Frozen in geologic time, proxies help to solve the puzzle of Holocene and Pleistocene temperature fluctuations.

Besides reading temperature, rainfall, and atmospheric composition by proxy along with scrutinizing historical records of weather, climate scientists build computational models. They program the models with variables that mimic interactions among the earth, sea, air, life, and sun in the real world, and then ask, "Does the virtual world reproduce the decadal to millennial variations recorded by temperature proxies? If not, why not?"

Maybe the ice cores, seafloor sediments, and coral reefs do not adequately sample the earth's actual climate history as well as the modelers are able to model it. Maybe the variables programmed into the model misrepresent the real earth in some unexpected way. Even as confidence in modeling grows, uncertainty in the science remains.

How climate varies on different scales presents the key challenge to forecasting the earth's future, to factoring out the background noise from the signal of human-induced warming. Ideally, climate science reconciles two realms on a time scale that matters to human society: paleoclimate temperature data and rigorous computational modeling. Climate scientists couple the remains of life to the laws of physics and chemistry in order to forecast the global future. The stories microfossils and tree rings tell and the laws physics obey must amicably wed.

When they do, they bear progeny with worrisome characteristics. Data from seventeen state-of-the-art climate models have converged on the same conclusion: the risk of multi-decadal drought in the southwestern United States and the Great Plains will rise in the next century. Conditions may become worse than recorded for the driest centuries of the Medieval Climate Anomaly, 1100–1300 CE.[22]

Reconciling the clans of fossil-finding and model-building climate scientists makes climate science irreducible to a single scientific method in any meaningful sense. Their methods must match the very different demands of measuring and modeling physical and chemical interactions on global scales, interpreting fossils and historical data, and computing the fluid dynamics of the atmosphere. Climate science combines isotope chemistry with geologic process and the laws of physics with probabilistic reasoning. By such adaptations it responds to the demands characteristic of solving problems of climate change.

The Challenges of Medical Science

The expertise of the clinical researcher in medicine departs in a very different direction. Prescribing safe and effective medical treatments that minimize risk faces the challenge of human variation and unanticipated side effects from treatments. Metabolisms vary from person to person across one person's life-span, by gender, and sometimes by ethnicity. High-quality medical research adheres to ideals of sampling and randomization in order to take this variation into account, isolate causal factors of disease, and gauge the efficacy of treatments.

The methods of medical science adapt to the complexity of the human organism. Randomly sampled and randomly assigned clinical subjects remain blind to the actual nature of their treatments, trusting the researchers to first do no harm—or at least inform them of known risks, which must be voluntarily accepted.

Randomness as an essential feature of scientific study may sound counterintuitive to a public shackled by indoctrination in the scientific method. The role of randomness in clinical medical research is more than a footnote to the stereotype. Better statistical controls sometimes overturn previously trusted medical advice. Statistical methods predicated upon randomization are essential to reaching conclusions. The proper application of random design constrains uncertainty.

However united the kingdoms of climate and medical science may be by an empirical attitude and the goal of reducing uncertainty, from the perspective of what scientists actually do they differ. The sciences are plural, their powers and methods diverse.

Diversity reigns because research agendas differ widely. Climate scientists conduct experiments in virtual worlds; medical researchers run clinical trials with living bodies. These profound differences deserve careful scrutiny by science educators attempting to reform school science, and they require abandoning the quest to unify the sciences.

Disciplines match concepts to methods in meeting the diverse challenges of inquiry. The sciences are plural, as are their fundamental conceptions and trusted methods.

Diversity reigns because the sciences design different methods to suit different purposes. Diversity reigns because phenomena of interest vary widely. Conveying the nature of the sciences must not whitewash their diversity. Depicting diversity, not privileging unity, best conveys the natures of the sciences.

NOTES

1. John L. Rudolph, "An Inconvenient Truth about Science Education," *Teachers College Record* (February 9, 2007), http://www.tcrecord.org, ID Number: 13216 (accessed January 20, 2015).

2. John Schwartz, "West Virginia Revisits Science Standards," *The New York Times* (Tuesday, January 13, 2015): A17; and "West Virginia Withdraws Altered Climate Curriculum," *The New York Times* (Thursday, January 15, 2015): A19.

3. Joseph J. Schwab introduced the science-teaching community to this phrase in his seminal work, *The Teaching of Science as Enquiry* (Cambridge, MA: Harvard University Press, 1962).

4. National Research Council (NRC), *National Science Education Standards* (Washington, DC: National Academies Press, 1996).

5. An index to the scientific practices and crosscutting concepts and other components of the NGSS can be found at http://www.nextgenscience.org/next-generation-science-standards.

6. Michael J. Padilla, "The Science Process Skills," *Research Matters to the Science Teacher* no. 9004 (March 1, 1990), http://www.narst.org/publications/research/skill.cfm (accessed January 19, 2015).

7. NGSS Lead States, "High School Evidence Statements," *Next Generation Science Standards: For States, by States* (Washington, DC: National Academies Press, 2013), http://www.nextgenscience.org/ngss-high-school-evidence-statements (accessed January 19, 2015).

8. NGSS Lead States, "High School Evidence Statements," *Next Generation Science Standards: For States, by States* (Washington, DC: National Academies Press, 2013): HS-ESS2-17, http://www.nextgenscience.org/ngss-high-school-evidence-statements (accessed January 19, 2015).

9. Performance Expectations consistent with the NGSS were created using the elements found in the standards' precursor document: National Research Council (NRC), *A Framework for K-12 Science Education: Practices, Crosscutting Concepts, and Core Ideas* (Washington, DC: National Academies Press, 2012).

10. A link to the Nature of Science understandings within the NGSS can be found at http://www.nextgenscience.org/next-generation-science-standards.

11. EQuIP rubrics may be accessed at http://www.achieve.org/EQuIP.

12. NGSS Lead States, "Progressions within the Next Generation Science Standards," in *Next Generation Science Standards: For States, by States* (Washington, DC: National Academies Press, 2013): appendix E, 2, http://www.nextgenscience.org/ngss-high-school-evidence-statements (accessed January 19, 2015).

13. NGSS Lead States, "Progressions within the Next Generation Science Standards," in *Next Generation Science Standards: For States, by States* (Washington, DC: National Academies Press, 2013): appendix E, 8, http://www.nextgenscience.org/ngss-high-school-evidence-statements (accessed January 19, 2015).

14. Randy L. Bell, *Teaching the Nature of Science through Process Skills: Activities for Grades 3-8* (Boston: Pearson/Allyn and Bacon, 2008).

15. National Research Council, *Teaching About Evolution and the Nature of Science* (Washington, DC: National Academies Press, 1998).

16. John Settlage, and Sherry A. Southerland, *Teaching Science to Every Child: Using Culture as a Starting Point* (New York: Routledge, 2007).

17. Earth Science Curriculum Project (ESCP), "Investigation 19-3: The Dinosaur Puzzle," in *Investigating the Earth: Laboratory Manual* (Preliminary Version; Boulder, CO: Johnson Publishing Co., 1964).

18. Charles R. Ault, Jr., and Jeff Dodick, "Tracking the Footprints Puzzle: The Problematical Persistence of Science-as-Process in Teaching the Nature and Culture of Science," *Science Education* 94 (2007): 1092–1122.

19. Commission on Science Education, *Science: A Process Approach* (Washington, DC: American Association for the Advancement of Science, 1965).

20. National Research Council (NRC), *National Science Education Standards* (Washington, DC: National Academies Press, 1996).

21. This sentence embodies the most important principle presented in this book and reflects the philosophy of Stephen E. Toulmin, who forcefully argued the need for each science to adapt to the demands characteristic of its own problems. His "characteristic demands of problem-solving" theme, as well as his attention to the "local, timely, and particular" phenomena that some disciplines address much more so than others, appears in *Cosmopolis* (New York: Macmillan/The Free Press, 1990).

22. Benjamin I. Cook, Toby R. Ault, and Jason E. Smerdon, "Unprecedented 21st Century Drought Risk in the American Southwest and Central Plains," *Science Advances* 1, no. 1, e1400082 (2015), http://advances.sciencemag.org/content/1/1/e1400082, doi: 10.1126/sciadv.1400082 (accessed February 17, 2015).

2

The Scientific Method from Cradle to Grave

The myth of the scientific method is the common ancestor of reform movements that anchor science teaching to habits of mind, process, inquiry, nature, practice, and crosscutting concepts. The myth persists as the never-ending debate over the proper number of steps in the method, and the myth's descendants have inherited this debate.

Analogously, reformers argue over what to list as essential processes, concepts, or practices common to all the sciences. Worry about the proper number of steps in a universal scientific method more than misses the educational mark; it skews public debate. Generic lists of broader abstractions do little better.

FOGGING THE ISSUE

Wyoming has witnessed a firestorm of debate about the guidelines for teaching climate change—and whether they prejudice the curriculum against fossil fuels. Despite support for the Next Generation Science Standards (NGSS) from energy giants Chevron and Exxon, Wyoming politics are sharply divided on the issue of adopting the new standards. The teachers of Goshen County, Wyoming, had already invested three years in curriculum development guided by the NGSS when the Wyoming legislature abruptly called for a halt.

As quoted in the *Casper Star-Tribune* and reported by the *New York Times*, Republican representative to the state legislature Matt Teeters objected to how the standards treated "global warming as settled science." Inculcating such science in the minds of Wyoming schoolchildren, he feared, would place the energy-export economy of Wyoming at risk.[1]

Undaunted, a Wyoming science teacher, adorned in a tie-dyed lab coat, responded by depressurizing a five-gallon jug containing a wisp of smoke and filled with moist air. The jug fogged as water vapor condensed on the smoke particles. The teacher had created a cloud-in-a-jar, a science demonstration made popular decades before national science standards and global warming controversies existed.

The climate debate gave new legs to the cloud-in-a-jar demo. The demo illustrated the basics of temperature, pressure, and phase change made important in the context of teaching about weather and climate. And it rebutted Representative Teeters by underscoring the importance of presenting evidence. In the teacher's words, "That's what the scientific method is all about."

The problem embedded in this perception extends beyond Wyoming into every science classroom in the nation. As pointed out in chapter 1, the problem is precisely about evidence—not just how it is presented, but how it was obtained, what counts (and does not count) as evidence, and what to do about uncertainty in light of evidence.

Different phenomena yield different kinds of evidence. Reasoning proceeds, in light of different kinds of evidence, in distinct ways, as starkly illustrated by the comparison of carbon chemistry to delta formation in the previous chapter. An appeal to "what the scientific method is all about" cannot settle arguments about climate change. Evidence matters, of course, but only in the context of reasoning guided by knowledge of concept and method.

A teacher must reach very far to make the creation of fog in a smoky jar count as evidence for global warming. The experiment does provide evidence that cloud formation depends upon the presence of condensation nuclei, in this case, smoke particles. What works in a jar by analogy works in the sky because the chemistry of water vapor condensation and evaporation remains constant.

Representative Teeters objected to the finished science in the NGSS that referred to, in effect, human responsibility for global warming and the need to understand climate science in order to make decisions to reduce the impact of global warming. The middle-school standard expressing this conclusion reads:

> Human activities, such as the release of greenhouse gases from burning fossil fuels, are major factors in the current rise in Earth's mean surface temperature (global warming). Reducing the level of climate change and reducing human vulnerability to whatever climate changes do occur depend on the understanding of climate science, engineering capabilities, and other kinds of knowledge, such as understanding of human behavior and on applying that knowledge wisely in decisions and activities.[2]

These propositions speak very vaguely to how the practices of climate science have adapted to the challenges of studying the earth's climate over time. The scientific practices and crosscutting concepts that bracket the middle-school global warming conclusions offer no guidance either. They simply reiterate the importance of modeling, using evidence, and determining causes whatever the context.

Representative Teeters has a point. Quite logically, he interprets the claim that human activities are a major contributor to global warming to mean a need to curtail certain activities such as fossil fuel extraction and consumption. He fears the consequences of such actions for his community.

Rebutting Representative Teeter with a vaguely conceived appeal to the authority of the scientific method falls as flat as the earth was once conceived to be. Representative Teeters has been mocked by his critics, but rhetorical mudslinging helps very little. It's as ineffective as the appeal to the authority of the scientific method.

Apparently, Representative Teeters welcomes debate about the evidence—he has his own agenda to advance and believes that evidence for potential economic harm due to regulating greenhouse gases is readily available. Interpreting the evidence of global warming, however, is a demanding task, irreducible to just the facts. Why to trust or doubt the evidence depends upon understanding the challenges and adaptations that are characteristic of a science.

The standards would do better to reflect the specific demands characteristic of climate science and how its methods of investigation and conceptions of phenomena have adapted to these demands. The science must meet the challenge, and schools ought to teach how it does so, without worrying about wrapping the science in the habits of mind or processes of inquiry common to all the sciences. That effort just restores an ineffectual notion of scientific method, as voiced by a Wyoming teacher, to prominence.

The teacher's homage to evidence drives no stake into the heart of global warming skepticism, though the cloud-in-a-jar does model cloud formation (not climate). In the end, his tired appeal to the scientific method cannot carry the debate, nor can the NGSS's generic list of practices clarify what climate scientists do. As a result, the politics of educational policy turns to shrill rhetoric.

Predicating school science on the overarching aim of teaching the scientific method has had a century or more to work its magic on the public mind. Disguised within the NGSS as practices, such teaching will continue to fail. The myth of the method is not up to the task of guiding the citizenry in productive, democratic debate.

The teacher in Wyoming spoke in terms all teachers and most citizens would readily accept, trusting that science done properly will eventually gather the evidence—the uncontested facts—needed to settle unsettled matters. At the very least, contesting facts makes for good scientific sport.

How did science education get started down this primrose path? Though not the sole instigator, John Dewey deserves much of the credit. A sense of "what the scientific method is all about" governed a Wyoming teacher's approach to instruction. No doubt, many teachers think similarly, believing their task is to democratize intellect in the form of teaching the scientific method to all students. Preparing citizens to reason scientifically and accept scientifically constructed arguments strikes a chord of consensus.

That aim underpins having all students learn science throughout at least a decade of their lives. It goes beyond promoting successful learning in science as a means of creating equal opportunity. Whenever a textbook begins with a description of the scientific method, whenever a public official invokes the scientific method as grounds for trusting a set of claims (or departure from it for doubting them), and whenever a teacher emphasizes an understanding of the scientific method, oversimplification triumphs. The aim is sound; teaching the scientific method simply cannot lead there.

DEWEY'S EARWORM

"Seventy-two, can do!" chanted the class of 1972 at the United States Air Force Academy as they marched each day to the dining hall. Americans are pragmatists drawn to the secular altar of what works—a can-do people. Problem-solvers earn respect in business, science, and society. The definition of dysfunctional government is inability to solve problems. In education, the federal Institute of Education Sciences maintains the What Works Clearinghouse. Americans appreciate a can-do attitude and value what works.

John Dewey ranks among the nation's premier philosophers of pragmatism, and to John Dewey goes much of the credit for characterizing science as a generic process. Historian and philosopher of science Michael R. Matthews points to the publication of Dewey's 1910 essay, "Science as Subject-Matter and Method," as a key moment in the history of teaching science as method.[3]

In this essay Dewey laments the teaching of science as little more than the accumulation of reliable information and not as habits of mind, a general scientific attitude, and especially "a method of thinking." For Dewey, failure to address these dimensions deterred students from pursuing science. Rote learning numbed minds and squashed interest. Authoritarian schooling reinforced "ductile, and docile" behavior[4] (Dewey's language)—less than ideal traits for democratic living.

Matthews stresses that Dewey believed civilization's future depended upon teaching "the scientific habit of mind." He further quotes Dewey's 1910 essay, "Scientific method . . . represents the only method of thinking that has proved fruitful in any subject."[5] Scholars from many disciplines have objected to Dewey's imperial reach. However, he primarily espoused an openness of mind to new evidence capable of solving puzzles and problems—of putting hypotheses to empirical tests and interpreting outcomes logically.

Deweyan science works to solve problems. It does so by constructing mental and physical artifacts useful in the conduct of inquiry. Ideas and methods—both engineered—adapt to the demands of the problem. The pluralistic, fluid vein embedded in Deweyan thought, with its depiction of concepts and methods as intelligent adaptations for problem-solving, remains viable and attractive.

As the twentieth century unfolded, Dewey's popularizers distilled his universalist views about scientific methodology into a short list of steps comprising the scientific method. Dewey opposed stultifying regimentation and its promotion of rote learning, as do many educators today. However, schooling has codified this thinking into the five (or four, or six, or eight, depending on the source) steps of the scientific method. "Hypothesize" and "test" are ubiquitous entries regardless of the number of steps.

Here's a synthesis representative of five-step procedures many teachers will recognize:

1. Form a question from observation.
2. Hypothesize an answer.
3. Design an experiment to test the hypothesis.
4. Accept or reject the hypothesis based on the analysis of the data.
5. Propose and test a new hypothesis.

Often step five is "Communicate the results" or "Draw conclusions." Proposing a new hypothesis is based on observing the results. The steps are therefore cyclical.

Mr. Spock, in *Star Trek VI: The Undiscovered Country*, in effect truncated these five into a single—and logical—principle: find the evidence for a theory that appears to fit the facts. Dewey was human, not Vulcan, but Deweyan logic and pragmatism, as simplified in the public mind, eerily resemble Mr. Spock's way of thinking.

Dewey never presumed to reduce the complementarity of thinking and doing to such a formula, although he came close, and John Rudolph has uncovered the history of this chapter of science education. Rudolph explains how Dewey, in his 1910 book *How We Think*, proposed a five-step heuristic that linked pragmatic thought to education.

Slightly paraphrased, Dewey's steps were: (1) feel the difficulty of a situation, (2) locate and define this difficulty, (3) suggest possible solutions, (4) examine the implications of these solutions, and (5) conduct further observation and experiment leading to a solution's acceptance or rejection.[6] The heuristic does have five steps, but they scarcely resemble the easily Googled stereotype.

Unsurprisingly, a lockstep method was anathema to Dewey. He courted flexibility, valued real-world problems, and distrusted compartmentalization of the curriculum. He was not the first to propose teaching a scientific method explicitly. Scientists for decades before him had advanced this notion in many ways. For example, even before Charles Darwin published *On the Origin of Species* (1859), his future bulldog, Thomas Huxley, claimed that unity among the sciences under the banner of experimental method merited prominence in university education. Huxley argued that learning science promoted a materialist worldview: the objective generation and justification of empirical claims restricted to natural causes. Huxley viewed scientific method as the proper antidote to religious superstition.[7]

In 1885 geologist Grove Karl Gilbert entitled his address to the American Society of Naturalists "The Inculcation of Scientific Method." Gilbert stressed scientific observation, distinguishing observation from inference and inductive reasoning from observations, and the importance of generating and testing hypotheses. For Gilbert, "The great investigator is primarily and preëminently the man who is rich in hypotheses."[8] Observe, infer, and then test hypotheses: this rhetoric in science education has entered its third century.

By Dewey's time, conversations about teaching scientific attitude and method to the burgeoning number of students entering secondary schools had many tentacles. After Dewey, a small number of steps came to stereotype scientific method. Since the 1960s, efforts to restore Dewey's sense of flexibility to solving scientific problems have repeatedly surfaced, as in the attention to practices in the NGSS. Calls for improving everyone's intellect by means of teaching science preceded and followed Dewey.

In these calls, oversimplified, one-size-fits-all thinking reemerges, stuck in the mind like an inquiry earworm. Set to the tune of "Twinkle, Twinkle, Little Star," it goes:

There are four steps to inquiry,
Can you list them all for me?
First, a question you shall ask.
Second, design a test or task.
Third, collect your data now,
Fourth, tell us what happened and how.[9]

Dogma's Baby Steps

Whether sung to the tune of "Twinkle, Twinkle, Little Star" or chanted rhythmically while jumping rope, "The Inquiry Jingle" readily reinforces memorizing the steps of stereotyped scientific method. It does replace "observe" with "ask a question," but it implies that a question is about something observed. Fourth- and fifth-graders who learn the jingle retain it for years, prepared to experiment ("design a test or task") whenever called upon to do so. The jingle does double duty as the spine of a lab report.

Elementary teachers value such jingles as mnemonic devices. The jingle helps kids remember the steps they will be held accountable for. Middle-school teachers appreciate hearing students effortlessly recite the steps in sixth-, seventh-, and eighth-grade science classes.

Teachers were asked to teach the four steps—a district policy. Conventional wisdom shaped step one: Ask a question. The step studiously avoids mention of "hypothesis" because "hypothesizing" was not required of young children in the assessment of inquiry skill. Most likely, the "anti-hypothesis" policy stemmed from overzealous application of child development theory; the logic of controlling variables could wait.

The jingle presupposes four steps but lists five—Dewey's original number. The last line ends with "and how." "And how" is a fifth step. Of course, this step calls for an explanation. It is not a procedural step, as are steps one through four. In slightly more formal jargon, despite the district's admonition to avoid expecting young children to hypothesize, these four steps become: state hypothesis, design test, collect data, report findings, and then offer an explanation of what happened.

The jingle served as a stalking horse of the steps of the scientific method. It was isometric to the domains of the state's rubric for assessing inquiry skill. Constrained by district leadership enamored of a generic conception of inquiry, teachers, led by the jingle's author, acted professionally and creatively, and they made science central to their elementary school. The jingle cast their ambitious efforts to teach science to children in acceptable jargon.

Nevertheless, ambiguity resides in each step. The inquiry process, as four steps set to rhyme, intentionally lacks context and content. Admirably, the jingle centers on "design a test or task." Scientists don't just do labs, they design experiments. Design depends upon the investment of knowledge and skill. Engaging students in the "design" of inquiry, not just doing experiments, is a modest step forward.

Short lists of the steps of the scientific method, whether converted to mnemonics or not, omit knowledge, describe no skills, yet end with a call for explanation. This approach fails to meet the "challenge of the match." No phenomenon has captured attention, no conception for posing a question or forming an explanation appears, and no investigative method adequate to the challenge yet exists. The jingle glosses over explanation in the "and how" coda. At least it reminds children of the parts to write up after doing an experiment.

The jingle speaks to universal process, not to particular inquiries. The song answers the call to prepare for assessment. Reciting the jingle demonstrates that children know what is expected of them by those setting the parameters of science teaching.

The jingle leaves a great deal of room for teaching science to children, but not with much direction. As an elementary-school rule of thumb, the jingle is okay. As an advance organizer for middle-school science, the four-step stereotype misleads more than it guides.

Children's science differs from disciplinary science, of course. Children have not learned to distinguish investigations into the natural world by domain. Children enter science lessons, especially in the primary grades, without disciplinary lenses (or blinders, to those averse to compartmentalizing knowledge when teaching young children). Art, science, and story permeate young learners' experiences when crafted by good teachers.

To the adult, young children's investigations may seem erratic and chaotic, messy and unfocused. As they explore phenomena, they begin to realize what is possible in the real world, a realization often accompanied by a host of questions. By stirring water with popsicle sticks soaked in food colorings, wisps of color spiral together and change, yellow and blue becoming green. Interest comes first, then discipline in pursuit of that interest.

Discipline in pursuit of interest is not the same as studying a discipline, such as microbiology. It's simply the precursor to thinking about the challenge of the match—what method of inquiry will be adequate for finding out more about something intriguing. The child's inquiry ought first to focus on encounters with the astonishing details of how the natural world works. It does not need to unfold in a prescribed set of steps—certainly not in invariant ones.

By the upper elementary grades the four-step jingle functions as a precursor to the worldview of science universal and triumphant: dogma's baby step. The march accelerates as the jingle reveals its true self: the state-mandated Inquiry Scoring Guide.

In essence, the jingle presages accountability imposed by the state, as in Oregon, for example. Oregon mandates that students in grades three and above complete inquiries scored with a four-part rubric: forming a question or hypothesis, designing an investigation, collecting and presenting data, and analyzing and interpreting the results.[10]

The same as the jingle, the rubric counts poorly. Collecting and presenting data are considered one step. Generic notions of observe, infer, predict, classify, measure, collect, and analyze data lurk behind the four steps. No content knowledge clouds the portrait of process embedded in the scoring guide. The rubric does add some scoring criteria by grade level but makes no distinctions about inquiry according to discipline.

Jingle Omissions

The stereotyped steps omit mention of "examine current understanding," in effect, step two of Dewey's original list and something good teachers commonly do. As a pragmatist, Dewey framed his list with an overarching concern for choosing methods deemed adequate to the challenge of the problem. Implicitly, step three's "proposing solutions" embodies this concern.

Dewey's step one has an essential affective dimension: feeling the difficulty of the situation and, by extension, the significance of the problem. Anomaly, puzzlement, importance—all figure in framing the inquiry.

Missing entirely from the rubric are "explore" and "debate." The jingle and rubric ignore the importance of an exploratory phase to children's learning in science prior to designing an investigation, and leave the impression that "interpreting the results" ends the affair. Exploration is essential to "forming a question," and debate within a community of investigators is crucial to "interpreting the results." That's where the integration of new knowledge with accepted understanding takes place—or causes conflict.

Clearly, good teaching may sandwich the four steps between exploratory and debate phases of instruction. However, in an era of high-stakes assessment and resulting standardization, the jingle and rubric point clearly in the direction of the preferred path. The myth of unified scientific method—in four-plus-one ambiguous steps—subverts the messy, child-centric diversity of exploratory learning, prematurely collapsing a child's struggle to match methods of inquiry to the challenges and obstacles presented by diverse phenomena.

The jingle is the arrogant rubric's heuristic surrogate, a tidy rule of thumb for training children to comply with official curriculum edicts. Ironically, though derived from Dewey's five steps of thought, the state's rubric's regimented tone reinvents the problem that Dewey set out to solve: compliant, docile, dutiful, ductile school behavior as the path to success.

To be honest, science lessons with young children sometimes unfold in perfect harmony with stereotypical scientific method. In order for this to happen, the all-encompassing rubric requires refinement. The four-step mantra is just too vague, too general, to do any good. The solution is to specify the scientific method more rigorously and more operationally. The price, of course, is to intensify the stereotype and corrupt the challenge of the match—to gloss over diversity of methods in favor of one that works pretty much the same all the time. But when it works, it really works—that's why it's so seductive. And "what works" has high appeal in American education.

CLASSICAL EXPERIMENTAL DESIGN

Representing the scientific method in school as a single-manipulated-variable experiment rescues the four-step mantra. It puts some flesh on very naked bones and certainly works well in a variety of contexts.

Consensus exists among science educators on one skill that all students ought to learn: Carry out an investigation by changing one variable while keeping everything else constant. This skill takes time and effort to learn—and knowing how to conduct a single-manipulated-variable experiment in one context does not mean knowing how to do so in all others. Children may invent this approach on their own when immersed in a problem for which it offers a compelling solution. More than likely, a teacher, parent, or mentor needs to train children how to conduct a single-manipulated-variable experiment.

For one thing, keeping everything else constant is quite tricky. What counts as "everything else"? If the effect of changing the length of a pendulum on its period is the experiment, what must be kept constant? The temperature of the room? The age of the experimenter? The color of the pendulum bob? The shape of the pendulum bob? The release point of the bob? Whether or not the bob gets a push when released? The calendar date? The force of gravity in the classroom? The name of the experimenter?

Clearly, "everything else" cannot literally mean "everything else" but only other factors that may matter. What may matter—what might influence how a pendulum swings—is a matter of judgment based upon knowledge and prior experience. An experiment cannot produce new knowledge out of uninformed method.

David Hawkins, a pioneer of the *Elementary Science Study*[11] of the 1960s, realized the importance of "messing about" with pendulums prior to conducting an experiment in the classical, single-manipulated-variable mold.[12] Watch and change things. Notice and wonder. Describe aloud. Swing pendulums side by side. Meet challenges:

make the bob go in a circle; use a long stick for a pendulum. Swing a pendulum underwater. Put a pendulum on a pulley and raise it while it swings.

Before designing an investigation, children should share descriptions and questions generated by simply messing about. They should be able to make what caught their attention and held their fascination happen again for others to see and talk about in their own words as part of an Exploration Symposium. The purpose of the symposium is to distill questions amenable to investigation and gain insight into what factors need to be held constant while one single thing is changed and an effect recorded. The changed factor is the manipulated variable, and the effect recorded is the responding variable (independent and dependent variables, respectively, in more formal terms). In designing an experiment, clarifying the outcome of interest—the response that seems puzzling—comes first. Messing about is prerequisite to being able to focus attention on what seems puzzling and in need of explanation.

From everyday knowledge, children will suspect that the calendar date is irrelevant, as well as the age of the experimenter. They will likely not worry about room temperature, shirt color, or hair length. Friction with the air is the primary force slowing down the bob's swings, and colder air is denser air, but these differences are subtle and probably too difficult to detect. From messing about, they likely have ideas about bob shape, pendulum length, release angle, and other factors that exert an effect on pendulum swinging. At last the time arrives to test a hypothesis by manipulating one variable while other, potentially influential ones are held constant.

The sophistication of the experimental process increases greatly at the moment children realize they can change which variable to manipulate. Variables should take turns as the manipulated one, producing multiple sets of data for comparison and interpretation.

The place to share findings about pendulum motion is at a Discovery Symposium. At this second symposium classmates can re-create features of their investigations, explain their designs, and then debate and question each other about just how well measurements were made and variables controlled. With trusted results in hand, they can pin down cause and effect: change the length or change the period but keep the bob shape constant. Even more talk about sharing efforts to replicate other students' findings at a third symposium is a good idea. By this time, children's talk can examine the value of rotating variables out of the "held constant" category (the controls) and into the manipulated one in a stepwise fashion.

Gravity is kept constant by doing the experiment in the same place. Surprisingly, to most people, the earth's gravity does vary geographically to a small degree. A very sensitive pendulum with length and all other attributes kept constant will vary its period ever so slightly in response to variation in the earth's gravitational field. Properly calibrated, a pendulum can measure gravity—a laboratory exercise sometimes pursued in a college-level earth science course. Such an exercise suggests the value of continuity in teaching the classical experimental method and of revisiting pendulum motion as increasingly sophisticated understanding develops.

The utility—the power—of classical experimental design is seductive. It works very well in the right contexts, and pendulum experiments are a perfect context for manipulating a single variable. Pendulum experiments render the law of gravity accessible to school labs, and the search for this law has cast a wide path through the history of science. Orbits, tides, and rockets obey gravity. Its existence gives weight.

Gravity is a fundamental force (electromagnetism is another), and nothing is more basic to physical science than conceptualizing force. Testing for the effects of forces acting on objects—such as balance, attraction, or repulsion—repeatedly admits to classical experimentation. Equally enticing is the accessibility of this method to novice learners. It works in the hands of teachers who may feel anxious about their own grasp of scientific knowledge.

Primary-grade students balance objects on beams. Upper-elementary students explore the advantages provided by simple machines in multiplying force. Middle schoolers learn that motions change in response to forces. Secondary students learn to represent force as a quantity with two properties—strength and direction—represented by an arrow. Forces spin objects, and spinning objects behave gyroscopically. As motions complicate, fascination grows.

From primary through secondary school, there is a great deal to learn from conducting classical experiments with pendulums and other phenomena. Reflecting on the investigative process itself is essential to forming a classical conception of experimentation. Classical experimental design works in a great many contexts, just not all.

The Black Squiggly Stuff

Setting up the aquarium in the blended second- and third-grade classroom has excited the children for several days. Now colorful guppies swim among the elodea plants, as numerous pond snails cling to the walls of the tank. The pump motor hums and the water filter bubbles. White sand an inch deep covers the bottom. The class is proud of their new watery world. It's day one. With luck, in time there will be more guppies and more snails, even if they are hard to count.

A few days pass. A young girl sits mesmerized, watching the guppies. There's something new in the aquarium: black squiggly stuff lying in strong contrast with the white sand at the bottom. The class gathers in a circle on the story-time rug.[13]

"Did somebody drop that black squiggly stuff into the aquarium?" asks the child. No one confesses, and all agree it appeared on its own.

"How might we figure out where it's coming from?" asks the teacher. Imaginative solutions follow.

"You could just watch the tank for a while," pops up as the most straightforward suggestion.

Since this is an opportunity for a science lesson, the teacher prompts the class by asking, "If the black squiggles came from inside the tank, what might have made them?" The prompt brings attention to the tank's components: fish, snails, plants, sand, and water. "What can we do to find out which one could be the source?" asks the teacher.

The teacher's question has set the stage for a single-manipulated-variable experiment. Quickly the class agrees what to do: put only fish in one tank, only plants in a second tank, and only snails in a third tank, then wait and watch to see which tank or tanks have black squiggly stuff on the white sand. They decide testing sand only is unnecessary.

Of course, a few worry that maybe the fish need the plants to survive.

Very soon afterward the solution reveals itself: the fish made it happen. Black squiggly stuff lies only on the white sand of the tank with the guppies. Could it be fish poop? The children's animated talk goes something like this:

"Ewww. There's even some hanging off one of the fish."
"It must be going to the bathroom."
"Do they pee in the water, too?"
"Yeah, and then they have to swim in it."
"Maybe we can find some snail poop."

The class has completed a classically designed experiment by manipulating the object in each tank. The responding variable was the presence or absence of black squiggly stuff. Ironically, the initial experimental design—"You could just watch"—was adequate to solving the problem, but that approach would have short-circuited the opportunity to introduce a carefully designed experiment.

The five-step method ends with proposing another hypothesis, or at least raising a new and related question. "Do snails poop, too?" is one example.

In this instance, the investigation took a different direction. Once the main aquarium was reconstituted and a few weeks had passed, the population of snails appeared to increase. The problem became, "How many snails are in the tank?"

The new question demanded a new method and the expertise of one of the older children, a third-grader skilled in multiplication. To the teacher, the new question was a math question, but to the children it was just another fish tank problem to solve. The class agreed that emptying all of the snails out of the tank and counting them would work, but might be rather hard on the snails. They agreed that another way should be tried.

Counting the snails on the front glass of the tank was possible, but snails on the back glass and much of the side glass were very hard to see. Plus, more were crawling on the sand and on the plants.

The teacher, who had previously taught introductory lessons on sampling and estimation based upon the *Elementary Science Study* (ESS) unit, "Peas and Particles," intervened with modest guidance. One student took the bait and proposed how to solve the problems of counting hidden snails and estimating how many snails were in the tank in all.

The student's method depended upon knowing how many snails were typically crawling over a square on the surface of the glass. Defining *typically* presented a small challenge. It required repeatedly and randomly selecting a square, then averaging the number of snails counted within it.

Finding how many of these squares were needed to cover the sides and bottom of the tank was the key to solving the problem—to totaling up all the snails. Adding up all the snails for all these squares, using the average or expected number of snails typically counted in a square, would estimate the snail population. The average number of snails per square times the number of squares needed to cover the inside of the tank would equal the snail population.

When the class assembled to hear the results, their conversation quickly turned to problems with the estimate. It did not count snails on plants. Maybe snails don't like to crawl on the sand. Maybe the spinner always landed on letters of squares that just happened to hold a lot of snails at that moment. Maybe most snails hid in the back away from light. Reasonable objections anchored in current understanding mingled with tentative acceptance of the estimate—a place to start.

Embryonic ideas of sampling, estimating, randomness, and likelihood of error swirled in their talk, all prompted by thinking about the adequacy of the match between the challenge of the problem and the method of its solution. Improving the method depended upon learning more about snail behavior and accepting the idea of counting snails in squares chosen randomly. Thinking and doing rowed in the same boat; knowledge and skill intertwined.

The third-grader both accepted the limitations of the method and believed it was good enough to figure out in the future whether the population of snails had changed. Perhaps just a change in the average number of snails in a randomly selected square held the answer to that question.

Later in the year this child wondered how many rice puffs were in a large box of rice puff cereal. There were too many to bother to count. Was there a shortcut? "Weigh and count a cup of puffs, measure how many cups of puffs fill the box, and multiply by the count" turned out to be the solution. (The third-grader who grasped sampling, estimating, and multiplying so well, by the way, became a math teacher as an adult.)

The path from the puzzle of the black squiggles to the problem of the snail population unfolded rather naturally. Children completed a single-manipulated-variable experiment that solved the first problem. Their design tested a series of hypotheses, rejecting all but one (the guppies do it).

Observing the aquarium shifted attention from fish to snails as the children defined a new problem. No hypothesis entered the stage this time around, but reflection on the adequacy of method intensified. Exploration and debate marked the beginning and end in both episodes of inquiry.

The following year the fourth-grade teacher brought a bucket of pond water to school. In short measure, two aquaria harbored a multitude of freshwater denizens. Then an interesting experiment began: one aquarium received an aerator, and the other remained stagnant. Both sat unobtrusively in the back of the room, and from time to time, children compared the tanks. They looked for changes in the populations of miniscule daphnia and cyclops (crustaceans) and posted sketches of creatures viewed under magnification. After several days they noted an explosion of undulating red worms (in retrospect, possibly midget larvae called "bloodworms") in the stagnant water.

Investigating environments built upon the previous year's study of populations. Children's knowledge functioned as a tool of inquiry.

The message is clear: language, knowledge, and purpose interact when children conduct inquiries about the natural world. The steps taken to find answers must make purposeful sense. Concepts are as critical to methods of investigation as are basic, scientific processes. No short list of stereotypical steps can capture this web of imaginative interactions.

Mis-educative teaching practices follow from ignoring this web, especially when children offer an unexpected answer to the test of a hypothesis. On the surface, they may seem to utter incorrect inferences. Careful scrutiny of their language and what it means, especially of how their words connect with shared experiences (which is the essence of an observation), may tell quite a different story, as in the mysterious case of the saltwater saucers.

The Case of the Saltwater Saucers

Often teachers depend on the expected result of an experiment to reinforce the meaning of a key concept: follow the procedure, observe the results, and label the results as an example of the new concept. In the next story, the targeted concept was "evaporation." The story illustrates how intending an outcome may hinder learning. Sometimes serendipity serves science quite well.

The children in the fourth grade were in the midst of a unit on mixtures and solutions. One way to tell them apart, they learned, was transparency. Solutions were transparent; mixtures were not.

Toward the end of the week, each student stirred spoonfuls of kosher salt into a plastic cup of water until none could be seen. They poured some of the salt solution onto shallow saucers and left these on a back table over the weekend.

By the next Monday afternoon the saucers looked very different. The teacher passed out magnifying lenses for the children to use in observing changes. There were no more shallow puddles in the saucers. Surprisingly, there was a marked rim of crystals where once the edge of the puddle had been. Children drew what they saw.

"What have you observed?" asked the teacher at one point. As the children shared their observations, she wrote their statements on the board. After a number of statements—each with an irreducible morsel of inference such as, "It's white now," "The liquid is gone," and "I can see the salt," one girl presciently added, "The water evaporated and left a solid solution."

The young girl communicated her observations using the scientific language of solutions in an unanticipated and unintended way. No one had taught her anything about "solid solutions." The expression was her own—crafted to capture what she had observed.

Her inventive language kept the connection between experience and observation intact. She did not observe water evaporating, since it's invisible as vapor, but the statement "the water evaporated," despite its inferential content, earned the teacher's acceptance. "The water evaporated" appeared on the board.

The teacher stopped writing at that point, signaling her detection of a misconception—an error in the girl's statement. "Look again—do you see that all of the water evaporated?" said the teacher. She intended the class to learn the proper meaning of evaporation. They were supposed to observe a dry saucer with crystals of salt.

The girl had seen, instead, a "solid solution" left by the water that had evaporated, so she repeated her answer. The teacher gently corrected her, explaining to the class, "All of the water has evaporated and left only the salt behind."

On the girl's desk sat her drawing. There was a square depicting a salt cube as seen under magnification. In the center of the square was a circle. The circle looked a bit like a drop of water. To the girl, imagining how to describe a liquid inside a crystal, this was the "solid solution" left behind after evaporation. To the teacher's way of correct scientific thinking, a solid solution was a non sequitur. Dry salt was the expected result, and evaporation meant leaving only the salt in the saucer.

But the center of the salt crystal certainly did look like a drop of liquid. Anyone who looked carefully would agree.

The girl's inventive language represented her best effort to communicate an anomalous observation. In fact, her phrase *is* proper, scientific jargon. The crust of the earth consists of enormous quantities of solid solutions, often referred to as rocks and minerals (not all of them, but many, such as the feldspars, which are a key component of granite and other rocks). Alloys of metal are solid solutions. Tooth fillings are solid solutions. Married people indicate their status with a solid solution—a ring.

The droplet inside the crystal was strange and unexpected—not a solid solution, but a solution trapped in a solid nonetheless.

No one reasonably expects a fourth-grade teacher's repertoire of science concepts to span multiple disciplines in detail from chemistry to mineralogy, though background sections in a curriculum might delve deeper than what is officially taught. However, failure to accept the girl's statement—and add it to others on the board—functioned as disapproval. She was not given the chance to help the others in the class, the teacher included, to see exactly what she was seeing—to achieve shared meaning for her "solid solution" phrase. For her, it meant that a drop of water appeared trapped in a crystal of salt, which is an "inclusion" in formal terminology.

Pressure from the teacher to be "correct" subverted a wonderful moment to truly examine the adequacy of prior knowledge for interpreting a novel experience. Learning an intended meaning discounted a serendipitous outcome and devalued very sharp thinking. Denying the validity of the expression "solid solution" had the effect of disconnecting a child from her experiences. Observations, she learned, were not to be trusted when leading away from the intended target of instruction.

Language is the stalking horse of meaning, and the clash between teacher and child over the language for interpreting the results of the experiment was troubling. An opportunity to examine an unexpected pattern was missed. "Is that water inside the little square? Or do crystals dry with curved and straight edges? If it's water, how did it get trapped on the inside? Will it stay there forever or leak out? Will little crystals form inside the one with the droplet, making a ring like the one in the saucer? Is the water in the cube still salty?"

Someone might add, "Maybe it's a trapped bubble of air." After all, that's what does happen in glacial ice as snow recrystallizes.

Updrops

Another day, another school: In a fifth-grade classroom, a ten-year-old stirs salt in cups of liquids, testing to find the ones that dissolve it. Left unstirred, salt in vegetable oil seems to become coated by a film and clump. The teacher suggests that children try adding oil to water to see what happens. One goes a step further. She drops a spoonful of salt onto the oil floating on water. The salt descends and crosses the oil-water boundary to come to a rest at the bottom of the cup. She watches, and then reports, "My salt does not disappear in the water. It makes updrops."

The salt sank to the bottom wrapped in a blob of oil. After a few moments, the buoyant oil slipped from the salt and rose to the oil-water boundary. Everyone wanted to see the updrops and then make their own. Her language captured her experience and communicated her findings just fine. Even the teacher was completely fascinated. The updrops resembled a lava lamp in miniature. Soon the class was testing what happens when a variety of substances were dropped into clear cups of layered water and oil: ink, food colorings, and powders.

The exploration mesmerized everyone; explanation eluded all. Children engaged in exuberant talk, often exclaiming, "Look at this!" Strange things happen at the oil-water boundary where forces of cohesion, adhesion, and surface tension interact—something to think about again and again in the future. Salt's properties and behavior in a novel context created an opportunity for speculative conversation true to the children's own expressive language. A new word, "updrops," functioned flawlessly, connecting language to experience and communicating with others.

If only "solid solution" had had a chance to work so well!

Abundant prompts for theorizing about what's actually going on when water and salt mix and then dry out eluded the fourth-grade class's attention. Theorizing was replaced by a regimented dash to the correct observation in the proper language. Teaching banished the imagery of solid solutions in favor of the proper scientific catechism. In reality, the phrase "solid solution" was just as accurate and useful as "updrops."

Theorizing, in the best sense of informed speculation empirically anchored, does not appear in the list of steps of the stereotypical scientific method. Anomalous observations—solid solutions and updrops—provide the perfect opportunity for children to talk science, that is, to theorize. It's okay for the mystery to remain a mystery. Children's science may, on many occasions, savor that message.

STEPS TO COLLEGE

A community college professor at a recent national meeting of geoscientists described how his entire department—professors from all of the sciences—aspired to teach the scientific method in the same fashion. In addition to aligning their syllabi, they wished to standardize lab-report writing in terms of an agreed-to statement of the scientific method.

They worried that learning to interpret the scientific method when presented as a different number of steps in different classes would confuse adult learners. First, they argued over the number of steps, ultimately settling upon six. All science courses for all science students reinforced this same set of steps, but without a jingle.

The science faculty objected to how every textbook in every science course had different steps, differently ordered steps, or a different number of steps. That variation ought to have cued them that their task was a fool's errand. Instead, they likened the variation to debate between "lumpers and splitters," referring to endpoints in how different researchers classify objects of interest, such as rocks. Needing to learn more than one way to interpret the scientific method was deemed something undesirable rather than the beginning of true insight into scientific diversity and how different methods overcome different kinds of obstacles to investigation.

Some textbooks limited the scientific method to just three steps. A committee at the community college worked to compile all the variants from many books serving multiple disciplines. A wizened departmental chair, in true Goldilocks style, had the privilege of making the final selection: not too many, not too few, but just right. That amounted to six steps:

1. Observation
2. Question
3. Hypothesis
4. Test/Experiment
5. Results
6. Conclusion

According to the inquiry jingle, they got the steps wrong: the children's four-step version placed "question" before "observation." Attempts to dispel the myth or stereotype of the scientific method continue to lead to generic, one-size-fits-all depictions of the sciences, disrespecting (or at least glossing over) the fundamental diversity of disciplinary concepts and methods.

At each junction in the expansion of science teaching—to educated elites at universities in the mid-nineteenth century, to the expanding population of secondary students in the early twentieth, to *Science for All Americans*[14] permeating the reforms at the twentieth century's end and into the twenty-first with the Next Generation Science Standards—reformers have lambasted the emphasis on memorization of facts and definitions, the inert right-answer drills dominating by means of direct instruction.

Dewey's dream of democratizing intellectual pursuit by elevating scientific method to universal status and Gilbert's eloquent characterization of the value of scientific hypothesizing have descended into stultifying regimentation. Huxley's faith in one indivisible, experimental method has led to public misunderstanding of how different fields conduct inquiry and wrestle with uncertainty.

The scientific method continues to stalk learners from cradle to grave, step by step, or step by step by step (depending on who's counting). Disguised within the NGSS as scientific practices, teaching the method endures and has done so for too long.

NOTES

1. Motoko Rich, "Science Standards Divide a State Built on Coal and Oil," *The New York Times* (Monday, May 19, 2014): A11, A12.

2. NGSS Lead States, "ESS3-D: Global Climate Change," in *Next Generation Science Standards: For States, by States* (Washington, DC: National Academies Press, 2013), http://www.nextgenscience.org/msess-wc-weather-climate (accessed January 20, 2015).

3. Michael R. Matthews, "Improving Science Education," in *Time for Science Education: How Teaching the History and Philosophy of Pendulum Motion Can Contribute to Science Literacy* (New York: Kluwer, 2000): 15. Matthews quotes from John Dewey, "Science as Subject-Matter and as Method," *Science* 31 (1910): 122. Reproduced in *Science & Education* 4 (1995): 391–398.

4. John Dewey, *The School and Society; The Child and the Curriculum* (Chicago: University of Chicago Press, 1990): 186. The phrase "ductile and docile" appeared in the original publication of *The Child and the Curriculum* (Chicago: University of Chicago Press, 1902): 13.

5. John Dewey, "Science as Subject-Matter and as Method," *Science* 31 (1910): 127. Quoted in Michael R. Matthews, "Improving Science Education," in *Time for Science Education: How Teaching the History and Philosophy of Pendulum Motion Can Contribute to Science Literacy* (New York: Kluwer, 2000): 13.

6. John Rudolph, "Epistemology for the Masses: The Origins of 'The Scientific Method' in American Schools," *History of Education Quarterly* 45 (2005): 341–76.

7. Thomas H. Huxley, "On the Educational Value of the Natural History Sciences," in T. H. Huxley, *Science and Education* (New York: Collier and Son, 1901): 40–63.

8. Grove Karl Gilbert, "The Inculcation of the Scientific Method by Example, with an Illustration Drawn from the Quaternary Geology of Utah," *American Journal of Science* 31 (1886): 287.

9. Anonymous.

10. Oregon Department of Education, "Official Scientific Scoring Guide," http://www.ode.state.or.us/wma/teachlearn/testing/scoring/guides/2011-12/science_inquiry_gr6-8_eng.pdf (accessed January 20, 2015).

11. Emily Romney, *The Elementary Science Study [ESS]: A History* (Boston: The Educational Development Center [EDC], 1973).

12. David Hawkins, "Messing About in Science," *Science and Children* (February 1965).

13. The aquarium story combines the author's experiences from 1972 to 1975 teaching life science lessons from the *Science Curriculum Improvement Study* (Chicago: Rand McNally, 1970). For examples of similar investigations see Chester A. Lawson, "The Life Science Program of the *Science Curriculum Improvement Study*," *The American Biology Teacher* 29 (March, 1967): 185–90, http://www.jstor.org/stable/4441653 (accessed January 23, 2015). The classroom conversations were reconstructed and synthesized from memory and not from transcripts. They are similar to the dialogues reported by Lawson.

14. Project 2061 (American Association for the Advancement of Science), *Science for All Americans* (New York: Oxford University Press, 1990).

3

Old Wine in New Bottles

Science educators have persistently asked, "In what ways should learning the content of traditional subjects serve as the means to more general ends, such as understanding the nature of science (NOS) or the processes of scientific inquiry?" Acceptance of these ends creates the problem of reducing the role of disciplinary context. This loss has been a regrettable consequence of laudable efforts to advance the cause of teaching science to everyone.

The enduring adherence to "a process approach" obscures how conceptualization intertwines with methodology. Reducing diversely adapted methods to generic processes obscures the ineluctable tie between disciplinary concept and investigative technique.

In discrediting a simplified, universal scientific method, modern views emphasize the diversity of inquiry methods and explanatory ideals across disciplines. A discipline's concepts themselves, such as "extinction" and "geologic time," function as tools of inquiry in distinctive and productive ways.

Paleontologists, for example, reconstruct the behavior of extinct beasts from fossil footprints using an appropriate strategy: comparing fossil imprints with patterns created by living creatures. The tracks of fighting hippos laid down in African mud bear upon drawing inferences from dinosaur trackways exposed in Jurassic shales. The footprints of fighting hippos "running and scuffling" resemble the tracks left by the lumberings of massive sauropods. [1]

In place of experimentation, paleontologists reason by empirical analogy. Analogy to the limb structure and footprints of the flightless rhea illuminates how bipedal dinosaurs likely moved. Paleontologists are perfectly good scientists, but they commonly do not manipulate single variables in classically designed experiments using dinosaurs as subjects. They rely upon other methods to infer behavior from fossils and overcome the obstacle to observation of extinction.

OBSERVATIONS AND INFERENCES

The stereotyped steps of the scientific method place observation first. Observing continues its leading role in the process approach. Observing certainly permeates the process of investigation—it's not just a first step. Casual observations spark interest. Trained observers see things others miss. Making purposeful observations requires careful setup—an experiment, for example. The entire purpose of an investigation may be to make a critical observation. Making an observation can be a complex process itself, the return on an investment in understanding.

Observations, simplified by the rhetoric of scientific method and teaching science as process, differ from inferences. Scientific observations record events and objects and their properties free of wishful thinking and biased expectations. Inferences come later from reflection upon observations or by combining them. Sometimes inferences are cast as predictions. If testable, a prediction may wear the mantle of "hypothesis."

The preceding rhetoric was brought to you context-free. Any context presumes language, and language imposes boundaries among categories of thought. All observational statements are cast with language and thus laden with prior thinking. There is no way to make an observation statement independent of an existing framework of ideas. Using words comes with consequences—the impossibility of making absolutely pristine observations being among them.

Observing Sunsets

Consider the simple observation, "The sun set at 8:00 p.m. in the west." Implicitly, the observer is standing somewhere on the earth, and many observers with synchronized watches standing at the same location can verify the observation. However, has the sun truly "set"? "Set" is a useful metaphor for everyday communication but hardly a scientific observation. Perhaps the observation should be, "The earth rotated into a position such that the observer's horizon eclipsed the line of sight to the sun at 8:00 p.m., Eastern Standard Time, looking westward." Awkward, but true.

The investment in precise language, drawn from an astronomical model, intensifies the effort to observe accurately and eschew inference. However, additional concepts have entered the fold: observer's horizon and earth's rotation. Ideally, by specifying how to make an observation, all observers following the same procedure ought to achieve agreement. They can do so only if they have achieved shared meaning. It takes a set of ideas—the language of conceptualization—to frame an observation.

In isolation, a statement is neither an observation nor an inference. In context, it might be either. The difference resides in its derivation—how the statement was obtained. Equally important is the grasp of its meaning. With acceptance of the method of derivation and shared grasp of meaning, agreement is possible.

People can agree that some statements are observations and others inferences based upon observations. "The sun will set a few minutes later tomorrow" is an inferred prediction based upon the movement of the earth in its orbit. What to make, then, of the statement: "The part of the earth in shadow shifts through the seasons, exposing the poles to sunlight at six-month intervals." Is the statement an observation, an inference, a prediction, or an explanation?

In large measure this statement functions as an explanation. Days grow longer and shorter through the seasons because the earth's shadow migrates. The shadow migrates back and forth annually exposing and darkening the North and South Poles because the rotational axis of the earth points in a direction at a small angle to the plane of its orbit. It tilts. The tilted earth spins on its axis and orbits the sun.

The shadow alternately engulfs each pole, north and south, for six months at a time. That's a pretty clear observational statement. From the International Space Station, the migration of the earth's shadow—the portion of the globe in daytime and in darkness—is observable. The migration of the shadow can be inferred from the accepted model of the earth's motion in space with respect to the sun. Drawing upon a mental model of the earth in space enables a host of inferences.

No one today—apart, perhaps, from a dedicated flat-earther—looks at a sunset and says, "Well, tomorrow a new sun will be born to rise again in the east." The sun, except in the lands of the midnight sun, disappears each night. Few persons now believe that each night destroys the sun. A shared conception of the earth in space precludes that thought. There's no "new" sun.

Ponder for a moment the meaning of "new" moon. There are *Many Moons*, wrote James Thurber, whose character, Princess Lenore, just wanted just to be happy.[2] She asked for the moon, believing that it was made of gold and the size of her thumbnail. The jester made a moon of gold for her to wear as a necklace. Princess Lenore did not think it a fake. After all, a new moon appears in the sky when the old one goes away. Everybody talks that way.

Observing Footprints

One exercise rises Phoenix-like, decade after decade, for teaching the oft-emphasized distinction between observations and inferences: the Footprints Puzzle.

The Footprints Puzzle entered the science-teaching stage in the mid-1960s as an exercise in the *Earth Science Curriculum Project* (ESCP).[3] The text accompanying the original exercise explained that birds and dinosaurs had traits in common (a prescient statement in 1964—birds are now considered avian dinosaurs). Flightless rheas, for example, left three-toed footprints evocative of stealthy *Allosaurus*, the predator from the Jurassic age.

Subsequently, the Footprints Puzzle appeared in middle- and secondary-school science programs,[4,5] science-teaching books for teachers (mentioned in chapter 1), and even along the Cascade Trail at the Oregon Zoo. A *MAD Magazine* cartoon in the 1970s almost duplicated the Footprints Puzzle as well.[6]

Fossil footprints exist, posing genuine puzzles about the behavior of extinct behemoths. Stone tracks freeze dynamic gaits in time, and scientists disagree about their meaning. School science refashions the fossil footprints into its own image in order to train students to distinguish observation statements from inferential ones in the abstract, despite the arbitrariness of this distinction outside of context.

As Footprints Puzzle exercises evolved, context—and real science—became less and less important. The exercise taught students to think in terms of observations and inferences and never to confuse the two. Observations must not contain elements of inference; inferences must be warranted by the observations. No distinction has permeated teaching the nature of science or the process of science more deeply than that between observation and inference. The history of the Footprints Puzzle illustrates this point.[7]

The puzzle primarily depicts footprints derived from drawings made of dinosaur trackways along the Paluxy River in Texas by Roland T. Bird in the 1930s. These tracks were exhumed for display at the American Museum of Natural History; the ESCP original acknowledges that similar trackways exist in other states.[8] Bird inferred two sets of tracks: one set left by a migrating herd of elephantine, brontosaurid creatures and the second by predatory carnivores that resembled the raptors of *Jurassic Park* fame.

The tracks intertwine, and at one point a predator's tracks disappear. Bird inferred that the predator at that point had clawed onto the flank of its prey, sunk its teeth into its neck, and went for a short ride. He had inferred a kill site.

The Paluxy tracks do not admit to a single interpretation. Vertebrate paleontologist Martin Lockley rejected Bird's speculation. Lockley concluded that the smaller predator walked the same shore as the prey species but at a later time. None of the larger dinosaur's footprints suggested an attempt at escape: no flight, no fight. Bird's inferred attack appeared impossible given the mechanics and size of the brontosaurid's tail.

Along the shores of Lake Manyara in Tanzania, Lockley found a modern analogue for the fossil footprints of the Jurassic. He compared tracks laid down by "pugnacious" male hippos with those left by predators attacking prey. Real attacks displayed "running, slipping, and sliding."[9]

Although inspired by Bird's discovery, the tracks depicted by the ESCP authors, textbook writers, zoo exhibit directors, and special publications for teachers on NOS held very little resemblance to the originals. They lacked the flavor of the paleontological debate between Bird and Lockley.

Educators refashioned the tracks in keeping with the goal of making observation clearly separate from inference. Real science disappeared. The curriculum version censored complexity and dis-embedded the footprints from a particular, Jurassic problem. Authentic science fell victim to the rush to abstraction, a common consequence of the process-centric approach.

The rather innocuous school science puzzle consists of three panels: two sets of different footprints converging, the two sets circling, then one set departing. Lockley described the panels as a fantasy. Nowhere on earth had such evidence ever been found.

Predators and prey do not behave in the way the tracks suggested. In the sanitized version for school science, two animals approach, and one speeds up to attack its victim. Both run in circles. "The poor victim succumbs, perhaps to dizziness," wrote Lockley. He further wondered why the presumed prey walked along rather slowly then chose to run in a circle to escape its predator.[10]

The faux footprints prompt inauthentic inferences. The puzzle is fun and therefore warmly welcomed. There is no challenging or confusing content. It matters little whether the puzzle suggests dinosaur behavior from ancient times along ancient, muddy shorelines or animals encountering each other on a snowy winter's day. The Footprints Puzzle is not real science—it's a fake representative of a basic scientific process.

A decade after criticism of the process approach took hold, Norm Lederman, a leading proponent of teaching NOS, lamented how teaching science subjects often lacked meaningful purpose and context. Isolated content and inert propositions invited rote learning.

Lederman pointed out that memorizing definitions and factual findings failed to introduce students to understandings of the nature of scientific inquiry that ought to be widely shared among citizens in a democratic state. Lederman proposed teaching a conception of NOS that distinguished between observation and inference and extended this distinction into the realm of theories and laws. His view also emphasized the tentative nature of scientific knowledge.[11]

The Footprints Puzzle proved central to his message. In 1998 *Teaching About Evolution and the Nature of Science*, to which Lederman contributed, popularized the three-panel strategy for prompting observation and inference. Imaginative science students inferred predator-prey interaction from the tracks and also imagined more fanciful scenarios: scavenging by condors or juveniles riding piggyback on adult dinosaurs. Perhaps they are imitation footprints created to fool tourists.

Stripping away paleontological context altogether or suggesting snow as the substrate keeps the engine of imaginative inference running. Teachers appreciate the interaction with their students that accompanies the exercise. Many believe it offers the perfect opportunity for teaching their students to think with the discipline of a scientist. Martin Lockley admits that the exercise is enjoyable and does stir multiple hypotheses into competition. What's to criticize?

The Footprints Puzzle works well as creative fiction because it stimulates good classroom dialogue. But fiction it remains, and real science it omits. A stylized diagram replaces authentic artifacts. The exercise does not open a pathway for interest to pursue, disciplined by relevant imagery—of pugnacious hippos, for example. Very little limits the context of observation.

Lederman proposed making science content a means to a more abstract end: understanding the nature of science, with distinguishing observations from inferences a part of that understanding. Ideally, observations and inferences are about something interesting. However, in the Footprints Puzzle, true to the process approach, real-world knowledge plays second fiddle.

On the face of it, understanding the nature of science appears to be a step forward from making generic processes the object of study in science. The exercises for teaching NOS suggest otherwise.

Science educators quickly conflated understanding the nature of science with practice in the use of scientific processes—SAPA redux (see below). In 2008 Randy Bell, one of Lederman's protégés, authored the popular *Teaching the Nature of Science Using Process Skills* to encourage teachers to do just that. According to Bell, learning to compose and classify statements as observations or inferences directly teaches students NOS. His book includes the Footprints Puzzle with the dinosaurian context fully presented and children instructed to imagine themselves at a field research site in Alberta, Canada.[12]

As practicing scientists, children learn not to embellish their observations with unwarranted inferences. However, the use of language necessarily imposes implicit inferences (recall the sunset example). Given the title of the activity ("Trailing Fossil Tracks"), children naturally suppose feet made the tracks. Familiarity with bird feet gives the tracks an unmistakable forward direction. The teacher's guide fusses over distinguishing whether a child's statement "the creature had three toes" is an observation or an inference and suggests that the statement the "impressions" have "distinct appendages" is a proper observation.

Language necessarily imposes inferences. The word "impressions" implies the condition of the ground where the dinosaurs trod. The language assumes that impressions were made in a soft substrate, in this case mud, by a living creature. There's something fascinating to think about: Did the creatures walk on intertidal mud, a soggy riverbank, or a submerged lake bottom? Materials simulating actual trackways can stimulate interest in such questions. Sometimes clumps of mud fall from dino feet as they walk. Sometimes a tail drags; sometimes it floats. Stylized puzzles might spark these speculations, but lack any real potential to pursue them.

Teachers may point out that in the first panel, "the two sets of tracks appear to converge on a common point." Clearly, that observation contains an inference. Maybe the two sets of tracks are diverging from a common point. Observing a direction of travel infers movement and a front and back to the track. It's hard to look at the tracks without drawing upon prior knowledge of foot anatomy. Three-toed creatures sport their toes in the front. Yes, in some animals toes do point backward—on squirrel hind feet when descending a tree, for example.

Observations just do not disambiguate from inferences as neatly as Bell's book would have teachers believe. Pursuing process glosses over this truism. Context matters immensely.

Bell does add context. He reveals to the teacher that the tracks were laid down by *Albertosaurus* and *Edmontosaurus*. Of course, since the tracks in the puzzle are fanciful and unrealistic, there's no way to confirm or discount this information. He cautions teachers to be careful with this information and not lead students to believe that the puzzle has a single right answer. Any answer will do if cast in proper observation and inference form.

Bell was not alone in resurrecting process skills to higher purpose.

In 2007 John Settlage and Sherry Southerland integrated principles of multicultural education with science teaching in a book for prospective teachers, *Teaching Science to Every Child: Using Culture as a Starting Point.*[13] In their effort to uncover cultural aspects of science, they presented observing, inferring, and classifying as the foundational elements of scientific inquiry.

The culture of science, from their perspective, built upon a shared commitment to the processes of science. They advised teachers to have students practice the skills of observing and inferring as a starting point for appreciating the culture of science. The Footprints Puzzle—the tracks in the snow version—appeared in their book as an exercise for introducing young children to the culture of science while keeping content at bay.

Exercises that subtract context and content not only deplete the resources for making inferences but also misrepresent how scientists observe:

> Those who invoke the puzzle as a means to teach the distinction between observation and inference as a hallmark of "the" nature [or culture] of science tend to strip it of authenticity and the rich array of resources for solving problems—for example, the use of analogy with extant bipedal birds or quadrupedal hippos to understand how extinct dinosaurs walked. The resources of rich imagery and relevant knowledge—content—become implicit inferences to weed out, not patterns of ideas to inspect.[14]

The issue is not that teaching the nature or culture of science necessarily misleads but that the process approach, as a fundamental principle derived from the scientific method for governing science teaching, subverts both paradigms. Both nature and culture approaches readily lend themselves to characterizing the diversity of the sciences.

Imagine truly applying a multicultural lens to the practice of cosmology in comparison to molecular biology, or to contrasting the concerns of clinical medical researchers with invertebrate paleontologists. Once the quest for unity subsides and the process approach has exhausted itself, science educators may find teaching the nature of science or the culture of science a liberating and productive experience, conducive to representing the true diversity among scientific methods and concepts, the tools of expertise.

Dwelling upon an oversimplified and decontextualized notion of observation and inference hinders such progress. As exemplified by the Footprints Puzzle, efforts to reform science teaching consistently fill new bottles with the same old wine—generic wine, not a distinctive Oregon Pinot Noir. These wine presses have been in operation for many years. As exemplified by the Footprints Puzzle, the process approach persists, threading its way through the culture and nature of science education.

FROM ARISTOTLE AND ROUSSEAU TO SAPA

The ancient Greek philosopher Aristotle believed in empirical observation as central to the formation of knowledge of the natural world. The search for the beginning point of teaching modern science to children as process, with observation front and center, could arguably extend to the Enlightenment era and the eighteenth-century work of Jean-Jacques Rousseau. In *Émile*, Rousseau wrote extensively about his pupil's efforts to observe a stick placed in a glass of water.[15] Émile noticed that the portion of the stick protruding from the surface of the water had inexplicably separated from the part of the stick submerged in the water.

Optics explains a glass of water's magical ability to break sticks (or submerged fingers) in two. Due to the refraction of light, the stick appeared broken but, of course, was not. The curved glass of water acted as a lens. The glass even magnified the submerged portion of the stick, as do curved glass jars that make olives look bigger than they are.

Rousseau realized that the child's language might communicate appropriate observations, but that the child could not properly interpret these observations using the concepts of optics. Offering a proper explanation, Rousseau believed, should not be made in haste. A child might utter adult words, but still not speak with adult meanings. Conversely, the child might express observation in language meaningful to the child, but not understood by the adult.[16]

Many observations were necessary—of the stick in and out of the water, of the appearance of the stick from different angles, of changes in appearance while adding water to the glass. The child needed to manipulate the shape of the glass, the stick's position, and the level of water—even look through the glass at a stick placed behind it. Experience richly informed by observation was essential to learning, argued Rousseau. Understanding took root in the rich soil of empirical experience. Understanding explanations took time and became meaningful once put to the task of making sense of puzzling observations.

Thousands of years after Aristotle, and in the centuries-long wake of the many issues Rousseau raised about child-centered learning, educators came to view "observing" as the most basic science process skill. Children needed to observe with their own eyes if they were to progress in learning science and be able to express themselves in precise statements unburdened by unwarranted inferences.

The institutionalization of science as process in modern times dates from the American Association for the Advancement of Science's (AAAS) program *Science: A Process Approach* (SAPA), published by the Xerox Corporation in the post-Sputnik era of reform.[17] Ever since SAPA, virtually all US textbooks, curriculum frameworks, and standards documents have included objectives phrased in terms of scientific (or "inquiry") processes.

Several of the same phrases popularized by SAPA a half century ago occur among the Next Generation Science Standards' (NGSS) list of scientific practices: modeling, interpreting data, and communicating.[18] Entries on the NGSS list correspond closely to SAPA, even when the words have changed. SAPA emphasized "formulating hypotheses, experimenting, controlling variables." The NGSS practices address "carrying out investigations." SAPA kept to tangible skills such as "defining variables operationally" (in terms of how to conduct an observation) while the NGSS's language remains more general: "engaging in argument from evidence."

OBSERVING A CANDLE

Observing is basic to science and ultimately inseparable from understanding. Perhaps no activity illustrates teaching science as basic processes better than observing a candle. This exercise has a long pedigree that began more than a century and a half ago with Michael Faraday's Christmas lectures.[19]

First a disclaimer: "There is no compelling reason to use open flames with elementary students," according to Terry Kwan and Juliana Texley, the authors of the National Science Teachers' *Exploring Safely.*[20] Adult supervision is essential. Alcohol burners are completely unacceptable, and hot plates are the preferred heat source.

Such activities using candles have permeated science teaching since the middle of the nineteenth century. Candle challenges, puzzles, and investigations abound on the Internet: candles made out of potatoes, candles burning "underwater," candles heating wires, candles heating a spoonful of sugar (as in the ESS "Mystery Powders" unit), and chilled candle smoke condensing into a film of water.

Wondering about "What's in a flame?" does require placing a flame at the center of attention; a hot plate will not do. "Observing a Candle" has served honorably as a lesson for introducing science process skills, and many prospective elementary teachers have encountered a burning candle in their "Methods of Teaching Science" class.

The venerable lesson on observing a candle distilled to its process basics typically commences as upper-elementary- or middle-school children or teachers-in-training receive a stubby candle. Observing begins before lighting: the candles are described, weighed, and measured. The teacher instructs the students to write down as many observations as they can about their candles. Lists grow. "The candle is white. Its shape is a cylinder. It feels slippery. The candle is 10 cm. tall. It's silent." They are prompted to use their senses, except that of taste.

Preliminary observations at an end point, the teacher moves through the room supervising the lighting of the candles. Observing continues as the students watch the stubby candles burn. "Look carefully at the top of the candle as you make your observations," the teacher says. "And above it, too. What changes do you observe? What colors do you see? Where are these colors? What is the shape of the flame? How does it change in appearance from top to bottom? Does the middle of the candle feel warm?" The learners work through a progression of questions similar to the ones Rousseau posed to Émile about the broken stick.

Responses come quickly: "There's some smoke. The flame is yellow and the wick is black. I see a glowing, red tip at the end of the wick. There's a little puddle on top of the candle. It's bulgy. The flame is very hot."

"Look very carefully at the bottom of the flame," prompts the teacher. Several children detect a bluish color beneath the yellow flame and notice a nearly clear space around the wick.

The candles extinguished, class discussion ensues. Observations are classified as either "qualitative" or "quantitative." The children consider the statement, "The candle will probably burn for more than an hour." The teacher points out that no one has observed the candle burning for that long. Hence the statement is not an observation but an inference.

The Wick Wicks

The lesson on observing comes to an end. One student asks, "What's the flame made of?" to which another quickly responds, "Fire, of course." The lesson on observing does not answer "What's burning?" Instead, the objective limits activity to recording sensory observations and simple measurements, then classifying them as quantitative or qualitative. No concepts are introduced about combustion; no causes are given for variation in flame color. No actions are taken to detect water vapor as a product of combustion. "The candle is burning" stands as a self-evident claim. "What's burning?"—a challenging question—remains unanswered.

In this case "burning" means the rapid oxidation of hydrocarbons. Vaporized wax is burning (hydrocarbon molecules at a high temperature are spontaneously combining with oxygen to yield water, new hydrocarbons, and carbon dioxide) as it turbulently mixes with air. These understandings are not part of the lesson and must not be pursued in haste. Developing the skill of observing is the objective, and headlong pursuit of this objective has sacrificed concern for understanding what's in a flame, what's happening as a flame burns, and why.

The teacher or a child may ask, "What's burning?" The potential answers are legion: the candle, the wax, the fire, the wick, the smoke, the air, the oxygen in the air. Observing exactly what is burning is very difficult to do—and to express. When trying to answer "What's burning?" concepts about chemical and physical change intersect with observing the flame. Observation alone fails. Stating observations requires language, and words have meanings.

Through the eighteenth century, chemists thought that burning meant capturing and releasing "phlogiston." As techniques improved for observing changes in mass during combustion, chemists found that the mass of products equaled the mass of the reactants. Mass was conserved even as fire burned.

Together with the discovery of the mix of gases comprising air, better techniques and an understanding of the conservation of matter displaced phlogiston theory. Observing progressed in tandem with understanding, each informing the other. Chemists realized that the substance known as "phlogiston" did not exist. That knowing changed the meaning of "burning." At one time "phlogiston" was the answer to "What's in a flame?" Burning released it.

A few students have written, "The wick is burning." Putting aside the issue of whether this statement is an observation or an inference, the "burning wick" is an excellent place to begin an open-ended science talk.[21] Attention focused on the wick has the potential to introduce puzzlement. "Does the wick get shorter? Does the flame touch the whole wick, part of the wick, or none of the wick? Is the burning at the wick, around the wick, or above the wick? How can the wick change so little while the candle burns so much? Is only the tip of the wick burning, or is the entire wick on fire?"

For a science talk to gather momentum, the teacher must have confidence in its direction *and* in the capacity of the students to engage in productive speculation even when they and the teacher do not really know the chemistry of a flame—a predicament shared with phlogiston theorists. Keep in mind that phlogiston theorizing did move science forward toward better conceptualization—and better observations—of burning.

Restricting instruction to skill-building is much safer than encouraging children to theorize on the basis of their own observations but with little knowledge of the phenomenon. Talking science—with partially formed ideas, creative analogies, limited prior knowledge, teacher confusion, and probably some misconceptions all swirling about—is messy. Process skills are tidy and basic, hence their seductive nature.

The wick "wicks," of course. It wicks the melted wax from the little pool at the top of the candle. Wax evaporates from the wick and mixes with air. Why does it ignite into flame? The match did it. But the question reappears: What made the match ignite? Does it burn in the same way as or a different way from the candle? What makes the flame flicker? Why are the colors different in different parts of the flame? What is the smoke made of? Will smoke burn? That one's testable: Extinguish the candle and hold a match in the stream of smoke just above the wick. The smoke ignites, the flame descends to the wick, and the candle reignites! The flame appears to drip down the column of smoke and sit on the candle's wick.

Remember, there is no need to make haste in finding the right answers. Speculation, wonder, and prior knowledge need time to simmer together. Comparisons spark questions: Gas stoves produce flames, for example. There's no wick and the color is bluer. Why? Pressure forces gas out the nozzle to mix it with air; therefore, no wicking is necessary. Composition and temperature change flame color.

Immediate right answers are not the point; organizing thoughts in response to observations matters more. Given context and prompted by anomalies such as observations about the wick, interest may build. That's one way to join children to learning science prompted by observing a candle. But in the process-based, "Observing a Burning Candle Lesson" the concern for learning science lies elsewhere.

Unless discussion moves beyond the prescribed lesson plan, no open-ended science talking unfolds. The focus on observation skill overshadows speculation about the process of burning. The aim is not to develop a conception of combustion related to the phases of matter (solid, liquid, gas, and plasma), the chemistry of oxidation, and the transfer of energy. There is little intention to pose puzzles and dilemmas as stepping-stones along a path to understanding how a candle works. The objective of the lesson is to acquire a basic skill of scientific process: observing. Simultaneously struggling to learn content only gets in the way, prompting confusion and frustration—or so argue process-objectives enthusiasts.

The lesson described conforms to the belief that science resides in technique, that is, in method, in the skilled process of observing. The phenomenon of candle burning is simply useful toward that end, not as a phenomenon of interest in its own right. According to the design of the process-centric curriculum, the context for learning the skill of observing mattered very little. A stick in a glass of water would do just as well. The skill of making quantitative and qualitative observations, and recognizing them in statement form free of inference, mattered most.

Faraday's Candle

Observing a burning candle stands as an exemplar of the process approach championed in the waning decades of the twentieth century and the waxing ones of the twenty-first. The nineteenth century featured a very different approach to observing a burning candle (admittedly, still practiced today in middle- and secondary-school physical science and chemistry classes):

> There is no better, there is no more open door by which you can enter into the study of natural philosophy than by considering the physical phenomena of a candle. There is not a law under which any part of this universe is governed which does not come into play, and is not touched upon, in these phenomena.[22]

So began Michael Faraday's mid-nineteenth-century lecture for a young audience. He compared a candle to a lamp and considered the source of tallow and wax. In his lectures he demonstrated simple experiments: tapping into the base of the flame with a glass tube and capturing "heavy vaporous substance" (the gaseous phase of wicked wax just prior to combusting); placing an airtight, glass jar over the candle and watching the flame diminish and extinguish; inserting a sheet of paper through the middle of the flame and producing a ring-shaped burn where the flame was hottest. "Where the heat and fuel come together," wrote Faraday. (Apparently, Faraday was not worried about the hazards of placing paper in candle flames. The trick was to place the paper in the flame only long enough for the hottest outer ring of flame to scorch it.)

Faraday demonstrated that burning produced water by condensing it on the undersurface of a cold vessel positioned over the burning candle. Water may suffocate a flame, but combustion's by-product is also water. Faraday connected the process of burning to human respiration and his analysis of the gases in the flame to the composition of the earth's atmosphere.

The flame was the object of interest and close observation helped thinking imbue this phenomenon with wonder. Deliberate manipulation of the phenomenon led to improved understanding of combustion. Faraday's candle flame ignited the interest of Victorian youth, and many chemistry teachers continue to use his simple experiments as demonstrations or laboratory exercises. They habitually put observation into a disciplinary context, a practice the SAPA approach often violated.

REBOTTLING SAPA

The SAPA curriculum introduced legions of teachers to very engaging and accessible activities. Its legacy is the ubiquity of process objectives. However, the process-centric approach—at times a virtual dogma—tended to devalue the messy realm of children's speculative thought and to detach process skills from disciplinary contexts.

The initial AAAS vision of unity across the sciences, defined as the processes of scientific inquiry, joined AAAS's second child, the themes and habits of mind common to all of the sciences, in 1990 as part of *Project 2061: Science for All Americans* (*Project 2061* also elaborated essential disciplinary understandings). The AAAS phrase "habits of mind" obviously echoes the language of Dewey's 1910 essay on science as method—a method for training the habits of mind.

SAPA reinvented the scientific method as fourteen processes for organizing units of instruction. The SAPA fourteen are broad and sweeping, yet have proven very amenable to classroom exercises, such as observing a burning candle. The basic SAPA skills are: observing, inferring, measuring, communicating, classifying, predicting. The integrated ones are: controlling variables, defining operationally, formulating hypotheses, interpreting data, experimenting, and formulating models.[23]

AAAS processes and themes have animated each subsequent cycle of reform in science teaching. AAAS language influenced the development of the original National Science Education Standards (NSES)[24] and persisted in the NGSS. Both the basic and integrated process skills remain well represented in the NGSS's list of scientific practices. Pressure to align curricula with these practices—to treat subjects as interchangeable parts for grasping the importance of investigating, modeling, interpreting data, and communicating—continues to obscure how, discipline by discipline, inquiry adapts to distinct challenges. Better understanding and improved methods advance in tandem.

For example, the chemist observes a flame while thinking about heat, phase change, and oxidation (or perhaps the escape of phlogiston) and trying to capture the products of burning. Invisible changes are an obstacle. The paleontologist excavates dinosaur footprints while wondering about the behavior of big game on the Serengeti. The fact of extinction is an obstacle. These obstacle-driven, distinct approaches and thoughts make the problems interesting and the science powerful.

There is unity, but it is unity within context: the integration of thinking and doing, a Deweyan ideal.

In the latter decades of the twentieth century, dissatisfaction with teaching science primarily as a means to learn discrete, generic processes of science took hold. Old wine needed a new bottle. The NSES fulfilled that role. Unfortunately, the content domains of the NSES resembled a treaty among Balkan States. Every paradigm of science teaching staked a claim: content connoisseurs, process aficionados, NOS philosophers, inquiry specialists, technology advocates, and science-in-society progressives.

The NSES arranged content in eight domains—side-by-side silos. Unfortunately, the NSES succeeded in making inquiry a content domain—actually, several domains—in their own right. In addition to the three traditional content domains (Life Science, Physical Science, and Earth and Space Science) that came to serve as the basis of paper-and-pencil and then computer-screen achievement tests, the NSES authors added Science as Inquiry, History and Nature of Science (NOS), Science in Personal and Social Perspectives, and Science and Technology. And, of course, there was one more: Unifying Concepts and Processes of Science. Among these unifiers are evidence, models, explanation, change, measurement, and a few big ideas such as evolution and equilibrium, form and function. Science as Inquiry (a separate domain!) distinguished between "abilities" for conducting inquiry and "understandings" about scientific inquiry.

The offspring of the process approach proved to be a multiheaded hydra, a monster that lurked in SAPA's subconscious.

In fashioning the next generation of the science standards, policy makers redoubled their efforts to fulfill the quest for unity: to distill the habits of mind common across all of the sciences, and predicate school science upon them. A simple tripod ("three dimensions") sufficed: crosscutting concepts, scientific practices, and disciplinary core ideas.

Unifying concepts found their home in the crosscutting concepts. Scientific practices inherited SAPA's list of processes. Once again, the standards parsed core ideas into three traditional subjects. The NGSS—rebottled themes and processes—stood ready for adoption by the states. Teachers were somewhat befuddled with trying to figure out how to translate inquiry into classroom practices while preparing students to test well on standards content.

The triumph of the unifiers has turned into a Pyrrhic victory: in the words of Diane Ravitch, testing "hijacked" the standards movement.[25] The three traditional content domains of the NSES claimed title to state-level, high-stakes assessment as well as grade-level scope and sequence planning. They continue in this role as the core disciplinary ideas of the NGSS.

Technology found itself somewhat hidden in the NSES. Not so in the NGSS, where engineering has coequal status with science. School curricula have yet to reflect the equal stature of engineering and science in the new national standards—a reform that would truly make a difference.

The NGSS also subsume a small set of understandings of the NOS—a somewhat hidden fourth dimension. NOS understandings permeate all fields. None is objectionable. Yet these understandings do not guide actual work, as does the idea of comparing rhea locomotion to how *Tyrannosaurus rex* strutted about.

The NGSS do reject teaching processes in isolation and proclaim that knowledge and skill function together. At the same time, they aspire to fulfill the quest to unify the sciences. Standards continue to frame science for all as if all science conformed to the same set of generic processes. If they truly did, then having children observe burning candles, manipulate lenses, and interpret fossils would pose the same obstacles to inquiry. They don't. Part of a flame is invisible. Lenses trick the eye. And extinction negates direct observation of past life. The old wine has a new bottle, but it tastes the same.

NOTES

1. Martin Lockley, "Chapter 13: Myths and Misconceptions," in *Tracking Dinosaurs* (New York: Cambridge University Press, 1991): 172–73.

2. James Thurber, and Louis Slobodkin (illustrator), *Many Moons* (New York: Harcourt Brace & Co., 1943).

3. Earth Science Curriculum Project (ESCP), "Investigation 19-3: The Dinosaur Puzzle," in *Investigating the Earth: Laboratory Manual* (Preliminary Version; Boulder, CO: Johnson Publishing Co., 1964).

4. Michael J. Padilla, "A Tale Told by Tracks," in *Prentice Hall Science Explorer: Animals* (Needham, MA: Prentice Hall): Student Edition, 26–27.

5. William H. Matthew III, Chalmer J. Roy, Robert E. Stevenson, Miles F. Harris, Dale T. Hesser, and William A. Dexter, *Investigating the Earth* (Boston: Houghton Mifflin Company, 1973).

6. Don Martin, "On the Trail with a Zoologist," in *Mad Magazine #132* (New York: E. C. Publications): 7–8. In Martin's comic two sets of tracks cross. The zoologist follows the smaller set of tracks only to discover a monstrous creature with tiny feet.

7. Charles R. Ault, Jr., and Jeff Dodick, "Tracking the Footprints Puzzle: The Problematical Persistence of Science-as-Process in Teaching the Nature and Culture of Science," *Science Education* 94 (2010):1092–1122.

8. Roland T. Bird, "A Dinosaur Walks into the Museum," *Natural History* 47, no. 2 (1941): 74–81.

9. Martin Lockley, "Chapter 13: Myths and Misconceptions," 169–73.

10. Ibid., 171.

11. Norman G. Lederman, "The State of Science Education: Subject Matter without Context," *Electronic Journal of Science Education* 3 (1998), http://wolfweb.unr.edu/homepage/jcannon/ejse/ejsev3n2.html (accessed February 4, 2010, from ERIC database).

12. Randy L. Bell, "Chapter 6: Trailing Fossil Tracks," in *Teaching the Nature of Science through Process Skills: Activities for Grades 3-8* (Boston: Pearson/Allyn and Bacon, 2008): 72–80.

13. John Settlage, and Sherry A. Southerland, *Teaching Science to Every Child: Using Culture as a Starting Point* (New York: Routledge, 2007).

14. Charles R. Ault, Jr., and Jeff Dodick, "Tracking the Footprints Puzzle: The Problematical Persistence of Science-as-Process in Teaching the Nature and Culture of Science," *Science Education* 94 (2010): 1106.

15. Jean-Jacques Rousseau, *Émile or On Education*, translated by Allan Bloom (New York: Basic Books, 1979): 204–6. Originally published in 1762.

16. *Émile*, 73–74.

17. American Association for the Advancement of Science (AAAS), *Science: A Process Approach* (Washington, DC: Xerox Education Division, 1967).

18. NGSS Lead States, *Next Generation Science Standards: For States, by States* (Washington, DC: National Academies Press, 2013). For current information about the NGSS, visit http://www.nextgenscience.org/.

19. Michael Faraday, *The Chemical History of a Candle* (Atlanta, GA: Cherokee Publishing Group, 1993). Originally published in 1861 under the title: *A Course of Six Lectures on the Chemical History of a Candle* given at the Royal Institution of London.

20. Terry Kwan, and Juliana Texley, *Exploring Safely: A Guide for Elementary Teachers* (Arlington, VA: NSTA Press, 2002): 60.

21. Karen Gallas, *Talking Their Way into Science* (New York: Teachers College Press, 2002). Gallas shares several cases of young children engaged in productive theorizing. She refers to these episodes as "science talks."

22. Michael Faraday, "A Candle: Sources of the Flame," Lecture I in *The Chemical History of a Candle*,1.

23. Michael J. Padilla, "The Science Process Skills," in *Research Matters to the Science Teacher* 9004 (March 1, 1990), http://www.narst.org/publications/research/skill.cfm (accessed January 19, 2015).

24. National Research Council, *National Science Education Standards* (Washington, DC: National Academies Press, 1996).

25. Diane Ravitch, "Hijacked! How the Standards Movement Turned into the Testing Movement," chapter 2 of *The Death and Life of the Great American School System* (Philadelphia: Basic Books, 2010).

4

The "Alphabets" from NSF

Decade by decade since the 1960s, the rhetoric of reform has displaced steps of the scientific method with new, more abstract phrases: scientific processes, scientific inquiry, the nature of science, scientific habits of mind, and scientific practices. Each phrase implies the existence of a singular science. Teaching for understanding of the unifying attributes of the sciences was expected to overcome rote learning. Thinking needed direction; drilling for right answers helped very little.

Whether in pursuit of the one true nature or the common culture of science, science education seems unable to escape from the confines of the scientific method and the process approach. Nature, culture, inquiry—and now practice—inherit the early codification of steps in the scientific method and their elaboration as the fourteen skills in *Science: A Process Approach* (SAPA).

From the 1960s through the 1980s, the scales tipped away from content in favor of process in progressive circles of reform at the elementary level. For teachers lacking a strong science background, process provided a convenient approach, if not a crutch. Processes were seen as simple and easy to explain. This approach treated a small number of content-free skills as representative of the sciences, suggesting that mastery of these skills might enhance student learning in several different subjects.

From the 1980s through the 1990s, educational researchers shed doubt on the efficacy of the process approach. Children appeared to reason rather logically when concepts held meaning for them. Their ideas served as resources for solving problems. Children taught process skills in isolation seemed not to learn very meaningfully.[1]

Despite research indicating its shortcomings, the process approach abides. It discouraged mixing content knowledge with training in process skills. Proponents worried that students might become confused by the content, feel discouraged, and therefore lose sight of the process objective.

SCIS AND THE WHIRLYBIRD

In a SCIS exercise from the 1970s for teaching about hypothesis testing, sixth-grade students were challenged to define the variables influencing the rotational speed of a "Whirlybird."[2,3] The Whirlybird consisted of a wooden arm with holes. It spun on a central axis, with its spinning initiated by winding a thick rubber band, much like coiling a rubber band to spin the propeller of a balsa wood toy plane. The Whirlybird looked like an upscale version of a Tinker Toy contraption.

By placing rivets in the holes, the weight of the arm (and its moment of inertia) changed. Perhaps more importantly, the distance from the center of rotation to the bolts could be changed. Students could manipulate both bolts and coiling, then count the number of revolutions completed until the released arm came to rest.[4]

Whirlybird investigations gave no attention to the concepts of angular momentum or inertia, ideas appropriately presumed too difficult for the intended upper-elementary-grade audience. Instead, lessons emphasized isolating an independent (or "manipulated") variable and testing for the system's response (the dependent or "responding" variable). Developing the ideas of angular momentum and rotational inertia, or any intuitive precursors to these concepts, remained outside the lesson's purview. They were considered unnecessary complications that might obscure the logic of experimental design central to the process of scientific inquiry. Ideas of manipulated and responding variables, the same as their observation and inference kin, stood alone.

Experimenting with Whirlybirds presumably taught students the process of controlling variables, an integrated processes skill. Consistent with observing a candle, another basic process exercise, the focus on process overshadowed constructing a conception of the phenomenon.

Content only gets in the way when instruction elevates and isolates processes. Overvaluing process has become a problem.

WHAT TO VALUE?

Science teaching has long valued the process approach and honored scientific methods. "What to teach?" is, at heart, the question, "What to value?" Answering raises the questions, "To whom? In what contexts?" Valuing process may dominate, but social issues matter to citizens in the conduct of their lives, and theoretical conceptions matter to scientists in the design of their researches.

Valuing unity guides science teaching in one direction; valuing either the context of social issues or scientific theories departs in another, more diverse one.

Citizens do have practical needs that differ from the demands of rigorous scientific theorizing and hypothesis testing as practiced by researchers. Historian George DeBoer's *History of Ideas in Science Education*[5] describes how science educators had crafted general science courses to meet these needs by the 1920s as well as how courses in botany and zoology ultimately merged to become high-school biology.

Concern for teaching science of value to the lives of citizens in general has coalesced in different decades under the banners of Science Technology and Society (STS) and Issue-Oriented Science (IOS), codified in the original National Science Education Standards (NSES) as Science in Personal and Social Perspectives and Science and Technology—the harbinger of elevating engineering to coequal status with science in the Next Generation Science Standards (NGSS).

Citizens' practical needs and the theorizing of the sciences are often difficult values to reconcile. On one side of the divide, curriculum may place chemical knowledge in a value context, as in Chemical Education for Public Understanding Program (CEPUP).[6] In one CEPUP module, the citizens of Fruitvale face a groundwater contamination problem. Middle schoolers map out the dispersal plume using a pH testing simulation and consider what policies are needed to mitigate the problem.

On the other side of the divide, inquiry reigns over theoretical fiefdoms. For example, the *Introductory Physical Science* textbook ushers students through an array of experiments in order to arrive by means of inductive reasoning at fundamental ideas such as the kinetic-molecular theory of matter.[7]

Despite the tension between issues-oriented and theory-building curricula, hybrids of inquiry and issue approaches have been tried as well. In *Chemistry in the Community* (ChemCom),[8] secondary students role-play citizens disturbed by fish kills that may disrupt the local economy. Investigating the fish kill problem serves as a context for teaching basic chemistry. Students proceed to discover that the bends (nitrogen bubbles in the bloodstream) killed the fish, the ultimate culprit being a hydropower dam's super-aerated water release. The release plunged fish deep into the river. When they rose, their blood released nitrogen that had been absorbed under pressure.

Those who criticize the issue approach leading to an understanding of policy find that it compromises scientific depth; reaction mechanism and the structure of chemical bonds tend to receive less attention than in a traditional chemistry course. Critics of both traditional textbooks and inquiry approaches that are intended to reveal the theoretical structure of chemistry lament the lack of social context. Time is the tyrant; the hardest value decisions are often what not to teach.

"How science actually gets done!" repeatedly surfaces as the rallying cry in answer to "What to teach, what to value?" Turning student attention toward issues never disappears, nor does it remain ascendant for long. A physics teacher deciding to open the window into the history of women in physics finds that it shuts very quickly when testing on angular momentum looms and the need is felt to keep attention focused on basic, "value-free" curriculum (free from examination of the social and political implications of science).

The theoretical structures and the well-warranted claims of the sciences, along with the techniques of investigation that produced them, function well as guides for what to teach and what to test for. Having citizens develop different perspectives on issues that matter to their personal and social lives presents a very different agenda.

Learning science in a sociopolitical context may antagonize one constituency in the community while pleasing another. Try teaching about reintroducing wolves in Oregon, for example. Finally, progress in understanding social and personal dimensions to science presents intractable problems of comparing student achievement. Scoring solutions to angular momentum problems is easier than judging knowledge of how science has opened and closed doors to women.

And yet, if "process" recedes as the answer to "What to value?" something needs to take its place. Just because the conversation is difficult does not mean it lacks importance.

FLYING SQUIRRELS

An Oregon biology teacher, for example, teamed in the early 2000s with the Oregon Department of Fish and Wildlife (ODFW) to find out whether flying squirrels were establishing populations in a nearby park with an extensive stand of second-growth forest. The students took turns on weekend nights working with a wildlife biologist to set live traps and conduct a flying squirrel census. The ODFW, however, placed a stringent requirement on the trapping: any nonnative and hence invasive squirrels, such as the eastern fox squirrel, had to be euthanized.

Because of this rule, not all students were willing to participate, yet all were fascinated by the natural history of flying squirrels and surprised to learn of their neighborhood niche close to school.

The school served a low-income community. The flying squirrel project turned around attendance problems. The teacher found the experience to be the best in her career and hoped never to go back to her traditional biology curriculum. She found herself thriving in a "school within a school" structure.

"Ecological niche" is a perfectly good theoretical construct. It's in all the textbooks. Investigating the limiting factors in relation to niche is good science. However, flying squirrels do not figure on state tests, while many other biology concepts besides niche do. The squirrel-chasers included many at-risk, low-achieving youth from poor households.

Time spent on the ODFW project was time not spent moving ahead with other, traditional topics. Administrators became anxious that test scores—and Adequate Yearly Progress (AYP)—might suffer as a result. Eventually, the large high school's entire semiautonomous academy structure, that which gave the teacher permission to base school science on a community partnership, was scrapped.

At least for a time, secondary students engaged in animated conversations about which squirrels belonged in the nearby park. They cared about the squirrels, and caring, believed the teacher, was a more valuable outcome than the traditional scope and sequence of biology instruction. Most importantly, students learned real science by doing science that mattered in their social, political, and personal lives. Above all else, they learned caring.

CONCEPTUAL VALUE

Joseph Schwab's seminal essay, "The Conception of the Structure of a Discipline,"[9] did not flinch from acknowledging the complexities of disciplinary knowledge. He worried that the public fundamentally misunderstood how science gets done and that it gets done in different ways. Ironically, science educators accepted Schwab as one of the founders of teaching science as inquiry, then proceeded to characterize inquiry as a unified content domain in its own right in the original NSES.

Across decades of science-as-process, teaching scientific inquiry, stressing the nature of science, and revisiting the scientific method in numerous guises, lessons from his essay have been lost. Schwab valued the role of diverse conceptions. He forged a link between methods of inquiry and disciplinary structures, captured by his dictum, "On the conception, all else depends," as quoted below:

> A fresh line of scientific research has its origin not in objective facts alone but in a conception, a deliberate construction of the mind. On this conception, all else depends. It tells us what facts to look for in research. It tells us what meaning to assign to these facts.[10]

The conception directs attention and determines meaning. His simple statement expressed a very complex analysis of the modern sciences. Truths had meanings irreducible to the discovery of facts. The conception functioned as the preeminent tool of inquiry. Only in the context of a conception may methods be deemed adequate to a researcher's purpose.

Treating inquiry as a content domain of its own has lost sight of this principle—and even divorced specific processes of inquiry from their dependency on a particular conception. The methods for finding a Higgs boson differ dramatically from those for figuring out how the Colorado River carved the Grand Canyon.

Of course, both physicists and geologists value empirical results, but the nitty-gritty of what to look for, what counts as evidence, and what meaning to assign to empirical observations depends upon a conception, the deliberate invention of mind. Conceptions and their accompanying methods of investigation are what trigger fascination.

Schwab's students often felt that his ideas about the structure of a discipline were too complicated to justify the trouble of learning them. In effect, they asked—channeling Joe Friday (the dour-faced detective of the 1950s TV show and 1987 Tom Hanks film, *Dragnet*)—for "just the facts." Schwab explained that facts were just fine but what mattered most was the framework for interpreting the facts—a framework constantly subject to revision. Schwab wanted his students to value the framework and its tentativeness and appreciate the basis for its revision. He valued a commitment to improved understanding over knowledge of the right answers in the moment.

Schwab concluded that science teaching, in order to reflect the practices of the sciences properly, must invite students to engage in inquiry *and* discussion. The course of study must acknowledge the tentativeness of claims. He famously dismissed traditional teaching as just "a rhetoric of conclusions."[11]

Current characterizations of the nature of science (NOS) have revitalized several elements of Schwab's perspective: the focus on the tentativeness of scientific claims and the inertness of facts not put to higher purpose, for instance. Process, nature, inquiry, and practice enthusiasts all may validly claim Schwab in their intellectual pedigree. However, in their quest to unify, they gloss over his "On this conception, all else depends" dictum.

Schwab focused attention on the warrant for a new claim produced in the context of inquiry. The warrant made sense within this context—hence his call for inquiry as a context for learning. Well-warranted claims had to conform to explanatory ideals. An explanatory ideal in geologic inquiry, for example, is to make certain that events cohere in time. Determining sequence is essential to constructing historical arguments. If the dates are wrong, the explanation is probably wrong.

Calling upon inquiry as context does not mean calling inquiry a content.

Inquiry's methods of investigation match the corresponding conception. Inquiry adapts to the characteristics of the problem of interest. Inquiry adapts to the problem of counting flying squirrels very differently from determining the toxicity of water to fish. The former must contend with the complexity of competition among species, the latter with the difficulty testing for potential (and invisible) chemical toxins. The squirrel inquiry's success depends upon applying a conception of niche, carrying capacity, and limiting factors. Solving the fish kill problem turns upon an understanding of solutions, pressure, and gas laws.

Good teachers know and practice inquiry from the perspective of the "conception on which all else depends." Unfortunately, the quest to unify succeeds by discounting the role of conceptions.

Schwab's influences on curriculum theory and inquiry teaching are not easy to untangle. Admittedly, those who kept close to Schwab's vision for teaching the structures of disciplinary knowledge found difficulty in making science accessible to all students. The Schwab approach favored teaching science from the vantage of abstract, conceptual development. Though he advocated for the importance of public understanding of science, many might argue that the curricula resulting from his way of thinking proved too abstract or too closely aligned with disciplinary purposes for adolescent learners. Disciplinary structures serving research purposes need radical transformation to become psychologically accessible and interesting to novices.

Trapping flying squirrels has greater appeal than "warranting claims" and parsing theoretical constructs. Even if aligned with an inquiry text (e.g., from the Biological Sciences Curriculum Study), the squirrel study demands time for debating the ethics of euthanizing invasive species at the expense of climbing the ladder of ecological abstractions.

Schwab placed a high value on learning about the conduct of scientific research through an inquiry approach to teaching. John Dewey similarly espoused the value of learning science through inquiry. Dewey, accused of fostering the popular stereotypical list of steps in the scientific method, also urged teachers to make the practice of inquiry the key vehicle for teaching science. The "good" Dewey honored disciplinary diversity as resources for solving problems and extending children's experiences, a trail Schwab continued to blaze. The "bad" Dewey—or at least his naïve popularizers—left a legacy of oversimplified, universal scientific method.

The context of inquiry promises not only to demonstrate authentic scientific practices, but also to prompt appreciation of the results. Trapping squirrels to conduct a census of their population in a disturbed habitat certainly counts as an inquiry exemplar. The inquiry reveals how the parts fit together: how conceptions generate methods, how methods warrant claims. The dimension of social value—caring about habitat protection, debating animal control ethics—extends inquiry to a value context.

Schwab would have approved. The conduct of inquiry—pragmatically adapted to real-world problems—prepares citizens for lives of social responsibility and democratic living. Investigating flying squirrel populations in local habitats creates citizens concerned about biodiversity.

The squirrel project nested inquiry within three important value contexts: the value of a conception of niche, the value of sampling methods to solving a niche problem, and the value of solving the population/niche problem to species conservation. Equally importantly, the approach linked adolescents to adults with both scientific expertise and a role to play in the implementation of public policy. Adolescents participated as researchers and as citizens, a Deweyan and Schwabian ideal.

While Schwab looked to the dramatic changes in twentieth-century science and the need to incorporate them into school science, Dewey drew upon nineteenth-century thinking about the importance of direct experience to children's learning. Dewey advised having students reenact the experience of original inquiry through curriculum activities. For Schwab and Dewey inquiry-style learning captured both the nature of the sciences and promoted meaningful learning.

KEEPING AMERICA STRONG

In 1950 President Truman signed the legislation establishing the National Science Foundation. Its mission was to promote both research and education. Truman declared that improving the teaching of science was an imperative of national security. Soviet pioneering in space reinforced this message with the launch of the Sputnik satellite in 1957.

Post-Sputnik worries about the need for the United States to maintain a competitive edge in science and technology prompted dramatic investments in science education by the National Science Foundation (NSF) leading to the era of the "alphabets": the aforementioned SAPA, Physical Science Study Committee (PSSC), Biological Sciences Curriculum Study (BSCS), Science Curriculum Improvement Study (SCIS), Elementary Science Study (ESS), and the Earth Science Curriculum Project (ESCP), to name a few.

In 1963 Harvard psychologist Jerome Bruner challenged educational psychology to acknowledge "that the foundations of any subject may be taught to anybody at any age in some form."[12] This principle gave impetus to the design of inquiry-style curricula in several disciplines and reinforced the telling question: How should schools depict content and process in scientific inquiry?

Unfortunately, the distinction between process and content became a chasm. The dichotomy promoted by the "how to depict content and process" question acknowledged that content multiplied and changed constantly, seemingly without limit. At the same time, scientific processes, especially those of experimentation, remained more stable, thus offering an enduring way of organizing school science.

The content-process distinction glossed over Schwab's characterization of how the sciences warrant claims, guided by a conception that determines "what facts to look for in research . . . what meaning to assign to these facts." By the 1980s, process objectives had provided a science-for-all rationale and content objectives came to serve as test fodder. Issues, values, and conceptions proved less workable for organizing school science from the top—too messy, too diverse, and too complicated. Content-process instruction proved more popular.

In 1986 the Reagan administration issued a dire warning in the same vein as President Truman's clarion call for a federal role in science: the United States had become a nation at risk, according to an open letter to the American people. With astonishing hyperbole, *A Nation at Risk: The Imperative for Educational Reform*[13] concluded that the nation's enemies seemed to have orchestrated the shortcomings of American schools. Achievement compared poorly with other industrial nations, putting American ascendency, if not well-being, at risk.

Of course, American educators and the American public had created American education. The effect of *A Nation at Risk* was to undermine confidence in their accomplishments. Initiatives to define high standards for student learning and restore this confidence gathered momentum.

Science educators looked over their shoulders at mathematics teaching and its rather successful progress toward authoring curriculum standards. Initially, the National Science Teachers Association, through its *Scope, Sequence, & Coordination* (SS&C) project,[14] took the reins. Aimed at secondary schools, the project attributed poor US performance on international tests to, among many other factors, the ubiquity of the "layer cake curriculum."

Bill Aldridge, SS&C leader and NSTA executive director, observed that other nations taught several science subjects continuously across several years. The United

States, on the other hand, isolated subjects into a single-year format: biology in tenth grade, chemistry in eleventh, and physics in twelfth. Mathematical proficiency justified the layer cake approach: physics demanded higher levels of mathematics than biology.

The mathematical proficiency argument was, however, more self-fulfilling prophecy than good psychology of learning. Either physics or biology can be taught with more or less sophisticated mathematics—whether algebra, geometry, statistics, calculus, or computer modeling. As a language of science, mathematics appears everywhere and at every level.

SS&C proposed overhauling the cake by having each year of school from grade seven and above tackle a full slice of all the layers. The SS&C approach entailed new textbooks, new curricula, and new training for teachers. Some teachers wondered if lockers were big enough to hold several science subject textbooks at once. Others worried about the effect of backpacks loaded with so many books on student posture.

More seriously, in the SS&C vision, biology teachers and physics teachers would coordinate topics of study in order to reinforce learning from multiple disciplinary perspectives. For example, the study of eyesight invokes evolutionary ideas about the origin of the eye as well as concepts of optics and color from physics. Retinas turn photons into neural messages by means of chemistry. That statement is certainly an example of disciplinary density.

SS&C planning indeed held promise, but the scale of the enterprise and its alien feel in American secondary schools was its undoing. The layer cake curriculum squashed its detractors. The voices of physicists even coalesced around inverting the cake: "Physics first!"[15]

Rather directly, one of the three dimensions of the NGSS continues in the spirit of the SS&C: the progressions of disciplinary core ideas. Core ideas are the principle component of the NGSS for aligning grade-level curricula. They function to determine content for high-stakes testing and for preparing US students to compare favorably on international assessments of learning in science.

The SS&C was fundamentally a curriculum-design initiative, not a standards-creation effort. The National Science Foundation (NSF) passed the science standards baton from the NSTA to the National Research Council (NRC) of the National Academy of Sciences (NAS) in the 1990s on the heels of the publication of *Science for All Americans* (SFA) (under the auspices of Project 2061 [AAAS—the SAPA people]).[16] In 1993 AAAS's *Benchmarks for Science Literacy* joined the mix.[17]

Sorry; acronyms abound in this culture.

For a time three visions for improving science teaching competed: NSTA's anti–layer cake model, the NRC's burgeoning standards, and the AAAS's literacy benchmarks. With the scientific method of John Dewey as Adam and the processes of scientific inquiry defined by the AAAS as Eve, they spawned a remarkable genealogy. In terms of visions for science education, the 1990s resembled Pleistocene hominid diversity; several, not one, species inhabited the landscape.

Benchmarks and standards do little to loosen the grip of traditional content on secondary science teaching—to make room for squirrel science. NGSS core progressions unfold traditional discipline by traditional discipline, suggesting limited prospects for radical diversification of science content from one region of the country to another (as would be the case if place-based teaching rose in prominence, a topic explored in Part 3 of this book). Where states adopt the NGSS, curriculum developers must align instructional materials with the themes and habits of mind common across the sciences—the subset of AAAS Benchmarks recast as crosscutting concepts—as well as scientific practices, the residue of process.

The alphabets of elementary science tagged along into the twenty-first century anchored to process and enriched with new versions of the *Science Curriculum Improvement Study* (SCIS and SCIS 3) and the *Full Option Science System* (FOSS kits).[18] SCIS adhered to Piagetian principles of stage-by-stage child development from sensory-motor, through concrete, to formal, logical thinking. SCIS developers organized grade-level topics hierarchically, with systems thinking—physical and living—as the grand organizer. Inquiry science captured the hearts and minds of reformist elementary-school science educators, best espoused in 1978 by Mary Budd Rowe in her *Teaching Science as Continuous Inquiry*.

Project 2061—Science for All Americans and its *Benchmarks for Science Literacy* also emphasized the importance of systems thinking. Systems thinking opened the door to engineering's coequal status with science in the NGSS.

FOSS modules built upon materials originally intended for special education, especially children with visual, auditory, and movement challenges: SAVI-SELPH kits[19] designed by the elves at the Lawrence Hall of Science in Berkeley, California (SAVI-SELPH—the best acronym in the alphabet tradition—stands for Science Activities for the Visually Impaired/Science Enrichment for Learners with Physical Handicaps). SAVI-SELPH kits featured easy to manipulate equipment and clever details such as taping washers to the bottom of plastic cups to reduce spills. Graduated cylinders came with markings in Braille.

Primary-grade teachers found the kits just as useful with mainstream children as with those in special education. Soon, their colleagues asked about extending SAVI-SELPH modules through sixth grade. The Lawrence Hall of Science responded with a "full option system." Several of the FOSS modules resembled the topical approach of the *Elementary Science Study* (ESS)[20] from the 1960s (e.g., *Batteries and Bulbs, Mosquitoes, Daytime Astronomy, Behavior of Mealworms, Peas and Particles*),[21] but with more deliberate attention to teaching processes of inquiry and eventually alignment with standards. FOSS remains popular.

Competing with FOSS and also funded by the NSF are the Smithsonian's and Carolina Biological Supply Company's Science and Technology Concepts program (STC). STC combines engineering with science, advertising "explicit engineering lessons in every unit."[22] For example, as part of studying the pendulum, STC adds the construction of an escapement using Tinker Toys to make a clock. STC recently expanded its program to serve secondary schools. The program features careful alignment with the NGSS scientific and engineering practices as well as crosscutting concepts.

Each try at something new rebounds to the tried and true. The alphabets of Truman's NSF, invigorated by *A Nation at Risk*'s call to arms, march on, reconfigured but subsumed by the same abstractions deemed common to all the sciences.

The task of selling science education's abstract superstructure—the process approach, the common themes and habits of mind, the unifying concepts, NOS, the crosscutting concepts—to parents and the public has never been easy. For them science is a gatekeeper subject—a doorway to higher education and employment. Scoring well on tests of achievement matters most and few care to invest the energy needed to grasp nuanced arguments about science in the abstract. For the public, just teaching the scientific method retains the needed progressive ring.

BACK TO SCHOOL NIGHT

Since the launch of the Soviet satellite Sputnik, many a parent has endured evenings hearing the earnest advocacy of new science programs. From winning the Cold War to maintaining economic competitiveness, America's future reputedly depended on keeping the science and engineering pipeline full. Schools have tried valiantly to teach parents what their children will gain by learning the alphabets from NSF.

One night in 1992, the district science curriculum specialist addressed the parent assembly on Back to School Night at Cliffside Elementary School:

"And then in the third and fourth grade blend, the children will be learning 'the subsystem and variable concept' using Whirlybirds. . . . Each subsystem consists of fundamental entities. . . . Children will learn to define variables operationally."

One parent interjected, "But what content and vocabulary will my child need to know?"

The principal responded, "To best serve your child, the district has adopted two science programs: a traditional textbook with content, and a kit program—SCIS—for teaching the scientific processes."

It was a wealthy district.

Virtually no one in the room had a clue as to what "fundamental entity" or "the subsystem and variable concept" meant. Most were more concerned about blending the grades and whether second-graders in the 1–2 blend would learn cursive as soon and as well as second-graders in a nonblended class. Others wanted to know if computers would be equitably distributed across each elementary school in the district.

An outspoken parent added, "I think it's very important to expose children to the language of science, even if they don't know what the terms really mean. Learning these concepts later will be easier if they are already familiar with the words."

A teacher answered, "Not to worry. We will teach the sixth-graders about orbits and gravity and why the gravity of the moon is less than the earth's—because it is so far away from the earth's gravity."

This vignette, which is an amalgamation of three school events experienced by the author, can be described by the phrase "what a mess."

The sixth-grade teacher obviously did not know the content he was expected to teach well. He held a misconception about gravity.

The principal had very little understanding of the nature of the SCIS program—it clearly develops conceptions of force and motion, organism and habitat, for example, organized within a framework topped by thinking about the interactions within a system. SCIS does address content.

The district science specialist spoke in an alien tongue. "Fundamental entity" was jargon from the unifying concepts and processes list that comprised the state framework for elementary science. "Subsystem and variable" pertained to the process component of the physical science unit in the SCIS program. Sharing these abstractions with the school's parent community communicated very little.

The parent expressed a popular myth: learning science depends primarily upon learning vocabulary.

Other teachers did not come to the rescue with an exciting description of the kinds of activities and their importance to understanding the natural world children would experience in the coming year.

What a mess. But the topic was science with children, not literacy or numeracy, the essential subjects. So no one really worried, and the evening went right on.

Thank goodness no one from the National Research Council had time-traveled from the end of the decade to explain the eight content domains of the national standards—and how "Unifying Concepts and Processes" differed from "Science as Inquiry." Nor had a fellow NRC time-traveler from 2014 appeared to explain how the next generation of standards had improved upon these ideas with "Crosscutting Concepts" and "Scientific Practices." The arc of reform in science education bends toward its own past.

OWL PELLETS

Led by a team of outstanding classroom science teachers, a large urban district in a West Coast city made ambitious plans to generate support among the parent community for teaching science as inquiry. They planned a series of workshops and selected an activity from FOSS using owl pellets to engage the parents of elementary-school children.

The inquiry they planned to share quite clearly integrated content and method. Conceptions of anatomy and the food chain shared the limelight.

Owl pellets are best obtained from a reputable supply house—one that has sterilized them. Spray bottles of water keep down dust and allergens. The pellets consist of the regurgitated, undigested remains of what barn owls, in this instance, have eaten. However, in the inquiry exercise, that's not the starting point. Participants are invited to speculate: What in the world is this thing? From time to time, someone is already familiar with owl pellets. The point is first to introduce the spirit of inquiry—of finding out through investigation, not by means of direct instruction. The instructor does not write a definition of *owl pellet* on the board to begin the lesson.

The activity in the classroom usually spans several afternoons. The pellets are dissected and the remains grouped. One way of grouping is by similar object: femurs, skulls, fur. Eventually the goal is to answer, "What did the owl eat?" The challenge becomes assembling skeletons of voles or other small creatures from their remains in the dissected pellet and gluing them on paper in their correct, relative positions. All a bit gruesome, but very engaging.

During an evening workshop with parents, the end point cannot be reached, but several skulls and femurs do emerge, prompting chatter and enthusiasm. The evening's activity pleased the workshop's planning team.

The district adopted the FOSS kit program and assigned two kits per grade level throughout the district. Children dissect owl pellets as part of the Human Body Module. Owl pellets with the skeletal remains of small rodents are more accessible and acceptable for dissection by children than are human cadavers.

The jump to human anatomy by analogy to rodent anatomy is not too difficult to make. However, answering "Why are they similar?" invites musings on evolutionary theory, a topic outside the bounds of the Human Body Module. The module plays it safe. The emphasis is on anatomy, but owl pellet activity can lead just as well in the direction of understanding food webs.

As an extension, the FOSS directions suggest dissecting sterilized chicken wings. The FOSS study of anatomy uses, in its words, the "scientific thinking processes to conduct investigations and build explanations: observing, communicating, comparing, and organizing." FOSS's STPs (Scientific Thinking Processes; another acronym can't hurt at this point) reflect the generic inquiry heritage.

STPs fail to make explicit the nature of comparative thinking in an anatomical context: how bones in the same relative position with respect to different bodies have features in common. This idea resides at the heart of homology, a fundamental conception in comparative anatomy. For example, the chicken wing is homologous to the human forearm, with substantial wrist and finger modifications. This approach to comparison, which applies to teeth, organs, and tissues as well, figures prominently in puzzling out evolutionary ancestry. It's the prime method for classifying the creatures of the world and boasts a pedigree that preceded Darwin.

Realistically, homology belongs to high-school biology. At the same time, FOSS has ventured into this conceptual territory and substituted a call to teach STP in the abstract for comparative anatomical reasoning. In this way, lesson after lesson, module after module, the ghost of the scientific method haunts opportunities to illustrate how (a) inquiry adapts to the nature of the problem and (b) how methods match concepts.

Owl pellets and chicken wings are tangible objects, the source of skeletal analogies to human anatomy. Thinking about the thinking in a particular context, adapted to purpose, focuses attention on the role of analogy. That's a sufficiently abstract goal, one that does not suffer from the effects of overgeneralizing scientific thinking processes.

A fourth-grade teacher in a bilingual classroom in the district took flight with the chicken wing homology. Her students examined limbs and wings, looking for attachment points of ligaments and discovering how they leveraged limb motion. Anatomical drawings, with labels in English and Japanese, wallpapered the room. Her students learned clearly how muscular-skeletal features functioned in chicken limbs.

"Structure and function" is a high-level conception tightly connected to generating investigative techniques—to suggesting what facts to look for and what meaning to assign to them, as Schwab would say. Anatomical structures have particular functions. The NGSS elevate "Structure and Function" to that status of a crosscutting concept—as useful in engineering as in biology. As a result, lessons tend to jump to an abstract, unifying endpoint before real problem-solving takes root.

Homologous structures may serve similar or different purposes. Often evolution puts old parts, in modified form, to new uses. The structure-function abstraction does have utility, but not until differentiated into disciplinary concepts and methods adapted to context does its power truly emerge.

In contrast to the wealthy school district's mess of subsystems and variables, fundamental entities, and dual adoptions in the portrait of Back to School Night and subsequent landforms lessons, the urban school district's owl pellets and human anatomy exercise communicated the spirit and style of inquiry teaching quite well.

Owl pellets, in the end, were surrogates for learning human anatomy—not owl behavior, not predator and prey relationships, not barn owl habitat and niche concepts, not homology subsumed by structure and function in the study of evolution. It's worth noting that different conceptions might guide owl pellet study and still integrate with the FOSS kit, especially if the pursuit of depth within well-defined contexts were to govern the curriculum rather than generalized STPs.

Despite success, attendance at the workshop was troublesome. Families living in poverty were underrepresented. The time devoted to leading the owl pellet activity with integrity subtracted from time available to examine the FOSS adoption more thoroughly. But inquiry did sell well. The evening communicated the spirit and style of children's science that the district was striving to achieve. The event was not the traditional Back to School Night, but a special event devoted solely to introducing the FOSS curriculum and inquiry teaching to parents.

The teachers who led the evening events for parents also organized summer inquiry workshops for teachers. At these summer workshops teachers did have a chance to learn more about comparative anatomy and ways to extend the human anatomy module.

ALPHABET SOUP

Learning elementary science through inquiry evolved from Whirlybirds to owl pellets as the NSF's alphabets made their lesson selections. Throughout this evolution,

SAPA's process approach kept its niche as a set of objectives encompassing all sciences when taught at the elementary level. NSF's acronyms swirl about, an alphabet soup stirred by inquiry's thumb, while few feel confident that the nation's approach to K–12 science has made America strong enough.

Schwab's recognition of naïveté in the public understanding of science launched a sophisticated era of secondary, inquiry-style curricula, especially post-Sputnik, that joined methods and concepts in common cause in high-school science classes. School science, however, oversimplified Schwab's dictum into a content-process dichotomy, and assessment followed suit, scoring inquiry with generic criteria and testing for content knowledge separately.

Even as science educators strive to portray the diversity of the sciences, the old one-size-fits-all bugaboo reasserts itself. The generic processes—observe, infer, predict, classify, control variables, collect and analyze data, model, communicate results—do little more than extend the list of stereotypical steps in the scientific method. Efforts to teach the nature of science or the culture of science have defined both in terms of the same list of processes introduced in the 1960s to overcome criticism of teaching the scientific method. Process-oriented teaching, whatever the justification (learning the nature, culture, or thinking of science), elevates the process skill to attention at the expense of relevant knowledge.

The elevation of process overlooks how knowledge itself is an important tool of inquiry—how all else depends upon a conception, a "deliberate invention of mind." The candle burns, the wick wicks, and the dutiful observer goes mindlessly on classifying observations as qualitative or quantitative. Having learned the catechism of observation, the student next tackles inference. Classification stands on deck and data collection waits its turn.

Tension exists between planning science instruction useful for citizens in general and recruiting scientists and engineers into the science and engineering pipeline. Making issues and science in social and personal lives central repeatedly stakes its claim, often engaging students in meaningful project work with community partners. Schools soon discover, however, that this dimension of teaching competes for time with understanding theoretical structures and mastering testable content. Issue and social dimensions of learning science, moreover, are very hard to evaluate.

Fortunately, skilled teachers do find ways to practice inquiry-style teaching in the context of solving community problems. They put disciplines to work while students work on mastering disciplinary ideas. Students immersed in solving community problems, however, have by design less time to pursue disciplinary knowledge, giving traditional curricula a potential advantage in preparing students for standardized achievement testing.

Standards encourage, of course, standardization and discourage curricular investment in the study of phenomena at the local and regional scale through various lenses. For example, prioritizing general knowledge of landforms and emphasizing the skills with classical experimentation irrespective of context marginalizes interpretation of the landscapes where students actually live. Testing for general content knowledge and separately scoring inquiry skills in the abstract reinforce this marginalization.

Deeply rooted in a false dichotomy, reducing science to process ignores how, as Schwab explained, a conception cobbles together thinking and doing. The process approach continues to metastasize. Despite the reduction of scientific universals to two categories plus NOS in the NGSS, the lists continue to enumerate twenty-three high-level abstractions common to all the sciences, the progeny of the alphabets mated to the standards. The lists, in effect, define scientific literacy independent of content, exacerbating the false content-process dichotomy.

Achieve, Inc. now champions the NGSS tripod: crosscutting concepts, scientific practices (engineering too), and progressions of disciplinary core ideas. Achieve markets them with a sophisticated rationale grounded in the research literature about learning science spanning epochs of reform. Unfortunately, the mindset for interpreting this research has changed very little over nearly a century; the dogma of unity prevails. The jargon of big ideas, scientific thinking, and process skills has seized the commanding heights of state level frameworks.

The NGSS simply gives new names to the apex-holding grand unifiers and leaves content traditionally arrayed. The promises Achieve champions belie the fundamental shortcomings of the enduring quest to unify the sciences. Unity in the standards purports to enable the fair comparison of learning outcomes student to student, group to group, and state to state. Fear exists that true diversity in curriculum design obstructs this agenda. However, by fomenting curriculum design that fails to do justice to the range of scientific enterprises, the quest to unify leads to the miseducation of American youth.

NOTES

1. Richard A. Duschl, Heidi A. Schweingruber, and Andrew W. Shouse, eds., *Taking Science to School: Learning and Teaching in Grades K-8* (Washington, DC: National Academy Press, 2007): 215.
2. Dorothy F. Gabel, *Introductory Science Skills* (Prospect Heights, IL: Waveland Press, 1984).
3. Robert Karplus, "Subsystems and Variables," in *Science Curriculum Improvement Study* (Chicago: Rand McNally, 1970).
4. Collin R. Boylan, and Douglas R. Hill, "The SCIS Whirlybird Revisited," *School Science and Mathematics* 86 (1986): 541–44.
5. George E. DeBoer, *A History of Ideas in Science Education: Implications for Practice* (New York: Teachers College Press, 1991).
6. Lawrence Hall of Science, "Fruitvale," in *Chemical Education for Public Understanding Program* (CEPUP; Menlo Park, CA: Addison-Wesley, 1991).
7. Uri Haber-Schaim, Peter Gendel, H. Graden Kirsksey, Harold A. Pratt, and Robert D. Stair, *Introductory Physical Science*, 10th ed. (Belmont, MA: Science Curriculum, Inc., 2010).

Originally developed and published by the Educational Development Center (Waltham, MA: EDC, 1967).

8. American Chemical Society, *ChemCom: Chemistry in the Community* (Dubuque, Iowa: Kendall/Hunt Publishing Co., 1988; New York: W. H. Freeman, 2002).

9. Joseph J. Schwab, "The Conception of the Structure of a Discipline," *The Educational Record* 43 (1962): 197–205.

10. Ibid., 199.

11. Joseph Schwab introduced the science-teaching community to this phrase in his seminal work, *The Teaching of Science as Enquiry* (Cambridge, MA: Harvard University Press, 1962).

12. Jerome S. Bruner, *The Process of Education* (New York: Vintage, 1963): 12.

13. The National Commission on Excellence in Education, *A Nation at Risk: The Imperative for Educational Reform* (Washington, DC: US Department of Education, 1983), http://eric.ed.gov/?id=ED226006 (accessed January 24, 2015).

14. National Science Teachers Association, *Scope, Sequence, and Coordination of Secondary School Science* (Washington, DC: National Science Teachers Association, 1992).

15. The Executive Board of the American Association of Physics Teachers, *Statement on Physics First* (College Park, MD: AAPT, 2002), http://www.aapt.org/Resources/policy/physicsfirst.cfm (accessed on January 24, 2015).

16. Project 2061 (American Association for the Advancement of Science), *Science for All Americans* (New York: Oxford University Press, 1990).

17. Project 2061 (American Association for the Advancement of Science), *Benchmarks for Science Literacy* (New York: Oxford University Press, 1993).

18. Delta Education (firm), Lawrence Hall of Science, University of California at Berkeley, *FOSS: Full Option Science System* (Nashua, NH: Delta Education, Inc., c. 2000).

19. Science Activities for the Visually Impaired; Science Enrichment for Learners with Physical Handicaps; Center for Multisensory Learning, *Introducing SAVI/SELPH* (Berkeley, CA: Center for Multisensory Learning, University of California, 1981).

20. Emily Romney, *The Elementary Science Study [ESS]: A History* (Boston: The Educational Development Center [EDC], 1973).

21. Elementary Science Study (Education Development Center [EDC]), *Elementary Science Study* (New York: Webster Division, McGraw-Hill Book Company, 1976; Nashua, NH: Delta Education, Inc., 1985). Developed and originally copyrighted by the EDC (1960–1973).

22. National Science Resources Center, Smithsonian Institution, "Experience STC," in *Science and Technology Concept*s (Burlington, NC: Carolina Biological Supply Co., 2014). Quotations are from the STC promotional brochure, http://www.pageturnpro.com/Carolina-Biological-Supply-Company/39073-The-STC-Program-(3rd-Edition)/index.html#1 (accessed January 26, 2015).

5

Abstract Liabilities

While generalized abstractions about science may appear to contribute to the task of comparing the outcomes of schooling, they actually have surprisingly little potential to do so. Vagueness plagues them, on the one hand; irrelevance to tangible courses of study, on the other. They specify too little within well-defined contexts, such as using homologies when doing comparative anatomy or extending sandpaper block and rubber-band analogies to real-world earthquakes.

Generic inquiry criteria misguide inquiry, forcing, for example, the study of landscape change and earthquake risk into the paradigm of classical experimentation. Seismologists certainly conduct careful laboratory experiments as well as field investigations. However, stereotyped science dishonors the diversity of their methods.

When the experimental method's central criterion, "controlling variables," leaps into prominence, "establishing temporal relationships," a procedure that functions analogously in interpretive science, limps in its wake. For sorting out the catastrophic effects of the most recent ice age or gauging the odds of the next Cascadia Subduction Zone earthquake, "controlling time" shapes the approach to inquiry more so than "controlling variables."

Admittedly, when diversity reigns, "What to Learn?" proves quite problematic from the perspective of management and accountability. This new set of problems, however, seems appealing in light of the liabilities of the standards movement. Zealous adoption of the Next Generation Science Standards (NGSS) seems fated to dependence on the wrong tools of assessment.

LIABILITIES OF THE STANDARDS MOVEMENT

Assessment practices drive a stake into the heart of diversity. State-level assessments require conformity from school district to school district; national ones, from state to state. Since its inception, the promoters of the standards movement have claimed that standards are not curriculum. Instead, they are supposed to do what uniform standards always do: facilitate measurement and, through measurement, improve quality. The standards approach to reform encourages competition that puts curriculum designs and instructional practices to the test. They converge as they compete.

In philosopher Nancy Cartwright's imagery, the sciences comprise a patchwork, not pyramid, of practices, and the sciences must adapt to an inherently "messy world." Cartwright warns, "The problem is that our beliefs about the structure of the world go hand-in-hand with the methodologies we adopt to study it. The worry is not so much that we will adopt the wrong images with which to represent the world, but rather that we will choose wrong tools with which to change it."[1]

Assessment is a primary means for changing educational practices. Choosing the tools for assessment must be done with care. Educational reformers seeking to change the world of science teaching are using a representation of scientific unity descended from over a century of teaching the scientific method. Belief in this aim has shaped methodologies for studying learning and results in questionable assessment practices. That's an inescapable liability of the standards, both the original flavor and those newly improved for tastes of the next generation. The wrong tools cannot change the world in the right ways.

The standards assume broad agreement with respect to "What to Learn?" When well-meaning people disagree on the particulars for student learning they often seek to build consensus by marching up a ladder of abstraction. Ideas about science and scientific literacy perched high on the ladder vary less than the details of subject matter. More general terms tend to find wider acceptance. Still, there are twenty-three separate practice, nature of science, and crosscutting summits (not counting core disciplinary ideas) in the NGSS. These summits are the points of consensus intended to overcome contention and promote scientific literacy for all Americans. They represent the habits of mind presumed common to all of the sciences. All means all.

Generic abstractions such as "observe quantitatively and qualitatively" and "infer tentatively" abound. These abstractions miss the target (meaningful learning in science) at the same time as they make a target of almost anything—cartoon sketches of impossible trackways of dinosaur footprints come to mind. Observing and inferring en route to tentative conclusions are inescapable features of sound reasoning. However, learning how the work of science gets done (and the more ambitious goal of learning how to do the work) requires learning about expertise grounded in disciplined practice. There's no escape by means of a ladder of abstractions.

Particular practices get the job done, not generic attributes. Particular practices embed inquiry within a conception. The universal dimensions of the NGSS promote the false opposite: the disembodiment of inquiry.

The falsehood of unity strands bodies of knowledge as traditional content domains. Anxiety fuels their embrace. In the standards, these domains coalesce around the topics expected to gain college admission. Subjects and topics that deviate from this consensus in response to local opportunity, student interest, or community partnership struggle for legitimacy. They thrive only if they are able to demonstrate a positive effect on achievement scores. They gain entry only if deemed exemplars of scientific ideals (the unity abstractions) and derivative of physical science, life science, and earth and space science. In other words, they must apply the same core ideas and illustrate the same universal features. Diversity scarcely has a chance.

Accepting fundamental differences among the sciences in terms of rhetoric, methodology, representations of data, and explanatory aims exposes the liabilities of the standards movement. Among its liabilities are the narrowing of the subject curriculum, the over abstraction of common elements among the sciences, the displacing of tools of the trades by generic abstractions, and the bifurcation of subject knowledge and inquiry skills in assessment.

From the perspective of honoring diversity, the alignment imperative ranks as an even more pernicious liability than achievement testing or inquiry skill assessment. The alignment imperative refers to scoring lessons, units, and curriculum for degree of conformity with the standards: the NGSS crosscutting concepts, scientific and engineering practices, and progressions of disciplinary core ideas. The standards are not the curriculum. However, curriculum design may not deviate from alignment with the standards—or funding evaporates and adoption prospects disappear. Together, assessment and alignment compel conformity. Conformity facilitates comparison.

As the epigraph from Gowin concludes, "The cycle is safe and therefore popular."

At the national level, two rubrics, Educators Evaluating the Quality of Instructional Products (EQuIP) and Primary Evaluation of Essential Criteria (PEEC), stand ready to evaluate compliance.[2]

Their scoring criteria check for the alignment of instructional materials with all three legs of the NGSS tripod. EQuIP and PEEC function to narrow and over-abstract the curriculum. These scores also reinforce displacement of the tools of the trades.

As an embodiment of valuing scientific inquiry skills promulgated by national standards, most any state framework will do. The story of Oregon's Inquiry Scoring Guide in the next section strongly illustrates the assessment liabilities endemic to the standards. Educators from many other states will recognize similar components in their official curriculum frameworks and assessment tools.

Assessment divorces high-stakes, state testing of knowledge in propositional form from task-centered, classroom assessments of student performance. Oregon's isolation of generic inquiry skills for assessment purposes promotes right-answer testing over problem-solving. Although hampered by the institutionalization of the quest for unity, the NGSS commitment to performance expectations as the backbone of assessment carries a modest prospect for reconnecting these bifurcated domains.

OREGON INQUIRY AND THE NGSS[3]

Each cycle of reform in K–12 science teaching has reconstituted the same abstract categories in slightly modified configurations. As a result, the NGSS's scientific practices and crosscutting concepts are simply the current iteration of the quest to teach the scientific method enriched with the themes and habits of mind espoused by Project 2061 and expanded as the fourteen Science: A Process Approach (SAPA) processes.

For Oregonians, to learn about scientific inquiry is to "understand science process concepts and skills that characterize the nature and practice of science."[4] This convoluted sentence combines several phrases of decade-by-decade jargon. It keeps "nature" and "practice" singular. Note that the reverse of this statement works—or fails—equally well: "understand the nature and practice of science that characterizes science process concepts and skills." Because these phrases encompass too much, they mean very little. The distinctions lack differences. Hence, their order is arbitrary.

In a related statement, Oregon defines "scientific inquiry" as "the investigation of the natural world by a systematic process that includes proposing a testable question or hypothesis and developing procedures for questioning, collecting, analyzing, and interpreting multiple forms of accurate and relevant data to produce justifiable evidence-based explanations and new explorations."[5] Whew.

When parsed the definition closely adheres to its five-step, scientific method precursor: observe, hypothesize, design an experiment to test the hypothesis, accept or reject the hypothesis based on the analysis of the data, propose and test a new hypothesis. Cycles of reform have encrusted the original; the scientific method has become "systematic."

This bureaucratic characterization of science rises to absurdity. For example, Oregon, after the fashion of many states, crosses life, physical, and earth and space science disciplines with nondisciplinary themes drawn from the quest for unity. The resulting matrix provides the Oregon Department of Education with the cells for reporting test scores.[6] Oregon labels the apex of this framework "Unifying Concepts and Practices."

The trail of abstractions for guiding assessment descends into a rabbit hole. In true Wonderland fashion many things are not as they first appear to be. Confusion grows as one abstraction references another with intensifying circularity. Each of these phrases had an origin in a different decade: "process" from the 1960s, "inquiry skills" from the 1970s, "nature" from the 1990s, and "practice" from the 2000s. "Standards" entered the lexicon in the 1980s. The concatenation of these paradigms makes the clear communication of expectations for learning in science to parents on a Back to School Night nearly impossible. They tend to reduce to, "So, you are teaching kids the scientific method?"

For example, Unifying Concepts and Practices are refined into the components Big Ideas and Science Processes. Categories that subsume big ideas must truly be grand. "Practices" unify "processes" out of the gate.

Understanding the "science process concepts and skills that characterize the nature and practice of science" falls under Science Processes. Thus, "Processes" subsumes "nature and practice." However, "practice" also sits on top unifying processes with "Big Ideas." Working downward from the apex, the progression is: practices processes process practice. Process has subsumed practice and vice versa. The score matrix is circular and symmetrical. Just when things seem to get smaller, everything grows bigger. "Curiouser and curiouser!" Alice would cry; she would feel at home.

Confused? Everything cannot be at the top at the same time—except in a science standards framework. The 1960s bequeathed the processes of science as an improvement over the scientific method. The twenty-first century's NGSS reengineered them as practices paired with crosscutting concepts, the identical twin of Big Ideas. Presumed progress is no more than an illusion masked by bureaucratic sophistry.

Oregon's Big Ideas component includes Interaction and Change. In the dialect of the NGSS, Stability and Change replaces Interaction and Change. Change is the constant. Oregon will need to update its table. Or perhaps it can strike out on its own and compose a novel category: Stability and Interaction. Equilibrium, anyone?

Is there no way out of this maze? Apparently not. Getting lost is evidence that something is amiss. States are reluctant to ask teachers to abandon frameworks assembled with their input at great expense of time and effort over many years. As a consequence the new encrusts the old.

The audience for science education is everybody, yet very few persons can navigate the labyrinth of unifying categories other than those who composed them and the bureaucrats charged with administering them. The situation is reminiscent of the emperor and the young boy who honestly appraised the situation and cried out, "The emperor has no clothes!"

A large, empty box appears where subjects cross with inquiry in the table that enumerates Oregon's eight categories for reporting testing results for science (see figure 5.1). There is no surprise that the blank space appears at the intersection of subjects and inquiry—an apt symbol of failure to wrestle with inquiry contextualized by subject in particular ways. The approach in Oregon, presumably representative of the way many states have institutionalized national standards, treats inquiry as a content domain in its own right.

The state expects teachers to assess inquiry skills at the classroom level using the Inquiry Scoring Guide. The scoring guide, modified slightly by grade level, is the same for all subjects. Teachers rate samples of student work on a six-point scale in four dimensions: "form a question or hypothesis, design an investigation, collect and present data, and analyze and interpret results."[7] The scoring guide is little more than the scientific method thinly disguised.

Adoption of the NGSS has caused Oregon to clarify its expectation for local assessment of student inquiry. The clarification emphasizes that the scientific practices and performance expectations in the new standards are similar to the previous four dimensions. All teachers need to do is require students to plan and conduct an investigation and complete a task scored with the existing four dimensions![8]

<table>
<tr><td colspan="2" rowspan="3">Score Reporting Categories (SRC 1-8)</td><td colspan="4">Unifying Concepts and Processes</td></tr>
<tr><td>Big Ideas</td><td>Science Processes</td><td>Scientific Inquiry (SRC 3)</td><td>Engineering Design (SRC 4)</td></tr>
<tr><td>Structure and Function (SRC 1)</td><td>Interaction and Change (SRC 2)</td><td colspan="2">Scientific Inquiry and Engineering Design (SRC 8)</td></tr>
<tr><td rowspan="3">Science Disciplines or Subjects</td><td>Physical Science (SRC 5)</td><td>Structure and Function in Physical Science</td><td>Interaction and Change in Physical Science</td><td colspan="2" rowspan="3"></td></tr>
<tr><td>Life Science (SRC 6)</td><td>Structure and Function in Life Science</td><td>Interaction and Change in Life Science</td></tr>
<tr><td>Earth Science (SRC 7)</td><td>Structure and Function in Earth and Space Science</td><td>Interaction and Change in Earth and Space Science</td></tr>
</table>

Figure 5.1.

An "official" scoring guide for Engineering Design has joined Oregon's one for science. It, too, identifies four dimensions with six-point scales for scoring samples of student work: "identifying and defining a problem to be solved, generating possible solutions, testing solutions and collecting data, analyzing and interpreting results." These dimensions parallel the ones for science. The acorn does not fall far from the tree.

As adoption of the NGSS proceeds, states ought to make efforts to specify disciplinary practices adapted to the demands characteristic of different problems. Oregon most certainly has not, and Oregon prides itself on following the national lead. As a result, the task falls to teachers to determine what tasks embody the actual concepts and methods disciplines depend upon.

In reality, the state lacks the expertise needed to specify practices discipline by discipline and the knowledge of problem characteristics from one field to another. Good curriculum development must attend to these issues—and the standards are not the curriculum. However, even the best curricula must align with the NGSS discipline-free characterizations of practices and concepts.

Standards first; curriculum afterward: Unfortunately, that is the national near-consensus. At some point dissatisfaction with standards-first dogma may gain momentum. Dissatisfaction begins by acknowledging how generic depictions of science eclipse sensitivity to its diversity. Generic scoring of student investigations exacerbates this problem.

Parents might take the lead by cross-examining presentations of Big Ideas, Scientific Processes, Crosscutting Concepts, Nature of Science, Common Habits of Mind, and Inquiry Practices by asking, "What is my child actually expected to learn and do with such knowledge?"

Teachers are in a more difficult position. Many have invested themselves heavily in making inquiry work as directed by standards, and all are expected to comply with district and state expectations. Among those expectations are that all science teaching will contribute to student appreciation of the practices and crosscutting concepts that unify the sciences. States demand thousands of hours from teachers through training workshops in order to align instruction and assessment with national frameworks. Alignment with grand, unifying themes makes this task unnecessarily difficult and wastes time—liabilities in addition to grand unification itself being suspect. A one-size rubric will never fit all inquiry designs.

Curriculum may readily embody real-world scientific expertise, of course. However, alignment requires that curriculum designers (and grant writers) demonstrate the unity, for instance, of how physicists figure out gravity, how ecologists determine food webs, and how climatologists forecast global warming. Short lists of practices and crosscutting concepts suggest that they do so in fundamentally similar ways and not as equally intelligent, yet very different adaptations of reason. Singularity, not plurality, rules.

Local nature centers, professional societies, and school leaders are certainly welcome to work out courses of study on the ecology of invasive species, the role of computer modeling in climate science, or even how art enhances the interpretation of earth history. These excursions, however, receive permission only if they don the cloak of unity and, perhaps more importantly, lead in the direction of higher achievement scores.

The standards are not the curriculum, but the standards determine what curriculum must accomplish. And an understanding of the scientific processes, however conceived, is what everybody needs. "If it does not align, you must not spend the time" becomes the administrator's mantra.

Oregon's Inquiry Scoring Guide provides an example of a wrong tool for assessment. Its application changes the world of instruction in unsavory ways. The guide ignores the importance of diversity among the disciplines. It is also a clear case of descent with modification from the scientific method original and therefore a reenactment of that myth.

A NATIONAL RUBRIC

EQuIP and PEEC are the counterparts to Oregon's scoring guide on the national stage. Together, they score lessons, units, and curriculum from all the sciences for alignment with the NGSS. Scores reflect how units, texts, and programs advance the crosscutting concepts, scientific practices, understandings of the nature of science (NOS), and progressions of core disciplinary ideas. Only the last category is subject-specific. The others reflect one-size-fits-all thinking.

Rubric-approved lessons call for opportunities to engage in the practices of science. Proficiency means successfully meeting the performance expectations that capture "three-dimensional learning" (the dimensions being disciplinary core ideas, scientific practices, and crosscutting concepts). Scoring stresses connections between disciplines either in terms of the crosscutting concepts or by virtue of shared core ideas.

Crosscut, core, and practice—ideally connected to the Common Core State Standards in several other subjects—dominate scoring. The two unifying dimensions displace high-level principles of inquiry embedded within disciplines. Core ideas such as "homology" or "frame of reference" should be drawn upon in setting performance expectations.

The performance expectations for assessing student grasp of core ideas are the most tangible and disciplinary-linked dimension of the NGSS. The original science standards made scientific inquiry a content domain in its own right (ditto for process and nature). The NGSS have steered clear of isolating practices and crosscuts, but nevertheless have perpetuated the generic approach. As a result, the abstract superstructure links loosely to all fields and tightly to none. It comes from the top by means of committee work, not the ferment of rich context.

Lacking direction from within-field principles of inquiry and governed instead by disembodied generalizations, progressions of core ideas traditional subject knowledge accumulates in propositional form: "The solar system is part of the Milky Way, which is one of many billions of galaxies," to use the NGSS's language. Core progressions consist of many such statements.[9]

Testing for comprehension of the unifying abstractions is troublesome, to say the least. Testing for knowledge of the components of the universe in propositional form (i.e., factual statements) is more straightforward. There are right answers.

Basing assessment on task performance encourages the integration of inquiry skill with knowledge. NGSS performance expectations promote such integration. Will states find their way through the thicket of problems surrounding the scoring of performance reliably, validly, and economically? The scale is daunting, and the siren song of testing for right answers is charming if for no other reason than it's cheaper.

Little prospect exists that states will ever be able to score progress in student grasp of abstract intangibles such as "modeling" (a practice) and "change and stability" (a crosscutting concept). In fact, they should not try to do so. A performance task must be cast with content in a context. Can educators assess student grasp of "modeling" by comparing performances on a modeling task in population ecology with the performances on one in climate science?

Success requires teasing out the contribution to performance made by specific content knowledge and domain-specific modeling skill from a general modeling proficiency, which may or may not exist. A similar argument applies to measuring growth in the grasp of "change and stability" even more forcefully. Setting generic ideas of science apart from context is fraught with peril.

Scoring lessons and programs in terms of EQuIP and PEEC rubrics has two negative effects: excluding disciplinary knowledge unrepresented in the core progressions and including generic ideas about science at the expense of disciplinary-specific characteristics. On all levels the rubrics narrow the range of sciences deemed appropriate for K–12 education and truncate depth of study within the fields they permit. Compliance fostered by EQuIP and PEEC scoring silences questions about fundamental aims. Those are already agreed to as the three dimensions. These are steep costs to bear for the imagined benefit of measuring and comparing student progress.

NOTES

1. Nancy Cartwright, *The Dappled World: A Study of the Boundaries of Science* (Cambridge, UK: Cambridge University Press, 1999): 12.

2. EQuIP rubrics may be accessed at http://www.achieve.org/EQuIP. A draft of PEEC was released in May 2015 at www.nextgenscience.org/ngss-peec-alignment.

3. This analysis first appeared in Charles R. Ault, Jr., "The Ghost Forests of Cascadia: How Valuing Geological Inquiry Puts Practice into Place," *Journal of Geoscience Education* 62 (2014):158–65. The table remained unchanged in response to the transition to the NGSS.

4. Oregon Department of Education (ODE), *Oregon Statewide Assessment Test Specifications and Blueprints: High School (2014–2016)*: 36. http://www.ode.state.or.us/wma/teachlearn/testing/dev/testspecs/asmtscitestspecshs_2014-2016.pdf (accessed January 26, 2015).

5. ODE, 24.

6. ODE, 40.

7. ODE, *Official State Scoring Guides*, http://www.ode.state.or.us/search/page/?id=32 (accessed January 27, 2015).

8. ODE, "Scientific Inquiry (SI) Local Performance Assessment Requirement and 2014 Oregon Science Standards (NGSS) Clarification." http://www.ode.state.or.us/news/announcements/announcement.aspx?ID=10369 (accessed January 27, 2015).

9. NGSS Lead States, "Earth Space Science Progression; Appendix E: Progressions Within the Next Generation Science Standards," *Next Generation Science Standards: For States, by States* (Washington, DC: National Academies Press, 2013): 2. http://www.nextgenscience.org/sites/ngss/files/Appendix%20E%20-%20Progressions%20within%20NGSS%20-%20052213.pdf (accessed January 27, 2015).

6

Cataclysms Outside the Classroom

Aligning all science teaching to the Next Generation Science Standards (NGSS) necessarily affects content selection. That's the point. National standards and state frameworks reinforce each other, shaping assessments and curriculum adoptions, and there are negative impacts to examine that follow from the assessment liabilities of the science standards movement.

Both state and national level unifiers distort science teaching. They tend to devalue studies responsive to children's spontaneous interests and curriculum anchored to a local story. Standards devalue the science most appealing to children. Standards give direction to elementary-school science not in terms of topics of spontaneous interest to children or investigations aimed at achieving intimate understandings of the local landscape, but in terms of abstract organizers.

These abstractions edit out context—children are expected to develop observation, inference, classification, data-gathering, and communication skills. Almost any subject will do in principle, and that's a problem solved by alignment with core ideas from traditional subjects. Standards assume the value of achieving common knowledge, measurable in similar fashion everywhere. Standardization, by definition, devalues eclectic mixes of topical study responsive to local opportunities to connect science to the personal lives of students.

Unifying standards and their generic, stereotypical depiction of scientific practices devalue topic work in elementary schools. They also let teachers off the hook by implying little need to know the content well. Many elementary teachers believe they can be good science teachers by immersing children in scientific processes. Conversely, they see themselves as not teaching science properly when emphasizing content. Reformers intend to reduce elementary teachers' levels of anxiety about science teaching. Being off the content hook helps, at the expense of helping children understand diverse topics in depth.

ICE AGE MELTWATER

A parent recently asked, "I understand fifth-graders will soon be required to take the statewide science exam. What are you teaching in fourth-grade science this year at Cliffside?"

"Landforms," came the response from the teacher. "During spring the fourth-graders will learn about landforms. They will thus be well prepared for state testing in the coming year."

The announcement of statewide testing of fifth-graders in science had focused district attention on what to cover.

And so the fourth-graders learned about landforms—mesas, deltas, shield volcanoes, and barrier islands—mostly from the textbook. Activity centered upon experimenting with a stream table. At the end of the unit, the teacher and classroom volunteers scored student-designed investigations of landforms created on the stream tables. No volcanoes erupted from the stream tables, but lots of sand moved around.

The school was located below the flank of a twin-coned, small, basaltic, Pleistocene, cinder volcano. Remnants of Ice Age flooding peppered the local landscape—a glacial erratic here, a landslide there, some lava beds stripped of soil, and sediment deposits grading from boulders to sand going upriver. The upriver finding of sediments posed a regional geologic puzzle solved not very long ago; they had derived from glacial meltwater on a continental scale flooding the local drainage basin.

The small volcano had pierced massive beds of flood basalts. Neighborhood road cuts exposed frozen, bubbly lava. Black cliffs sported hexagonal columns. In a nearby park, signs marked pits where bog iron had been mined circa 1900. In the surrounding county, dozens of small, extinct volcanoes dotted the landscape. In wet years landslides captured news headlines.

At the start of the study of landforms, no one at Cliffside Elementary knew how to interpret the remarkable local landscape, the product of volcanism and Ice Age meltwater.

At the end of the unit, no one knew how to interpret the local landscape any better.

The right-answer test, guaranteed to have questions about landforms, loomed on the horizon. The test would be more of an exercise in testing for reading proficiency than for an understanding of the nature of science or scientific habits of mind. No one expected to find any questions about how geoscientists had put together the story of the local landscape. State testing did not test inquiry proficiency; that was the classroom teacher's task.

Grasp of scientific inquiry was assessed by having each student complete an "inquiry work sample" to be scored by the teacher using the guide discussed previously. Each student first composed and then proposed to test a hypothesis using the stream table. The experiment had to include a manipulated variable and a responding one—the key players in classical design. For example, tilting the stream table manipulated the variable "gradient." In response, the stream might flow straighter or with more meandering. Manipulating gradient also changed the movement and deposition of sediment in the stream table. A single manipulated variable in a stream environment has several effects.

Students wrote up their reports following a careful four-part outline: asking the question and stating the hypothesis, designing the experiment, collecting and analyzing the data, communicating the results. The teacher scored each of these categories on the six-point scale, as she had been taught to do during in-service training. The report outline conformed to the elements of the inquiry jingle. That was not a coincidence; the inquiry jingle was invented to reinforce these elements, the skeleton of scientific method.

The stream table modeled landform construction and destruction due to the work of runoff. It represented the landscape in the abstract, investigated with the simplified method of manipulating a single variable. Scoring bumped abstraction to an even more general level: four dimensions of inquiry equally applicable in all contexts. School, district, state, and national efforts to improve the teaching of science in the elementary grades promoted each aspect of instruction. The teacher dutifully executed the landform lesson plans and inquiry work sample assessment. The test loomed, and landforms would be on it.

The local effects of the waning millennia of the Pleistocene Ice Age were less abstract: gorge-carving, land-sliding, shoreline- and cliff-etching, bedrock-scouring, soil-stripping/depositing, and remarkably anomalous upstream-facing deltas. The school sat on the floodplain where meltwater once cascaded in a volume equal to the combined discharge of all the major rivers of the world flowing today—the once suspect and now well-warranted geologic claim of J. Harlan Bretz.[1]

Gravel terraces from the Missoula Floods snaked streamside within easy walking distance of the school. A framework for interpreting these facts called for one that would assign meaning to the facts on the ground. They meant one thing: cataclysm.

The stream table model inadequately represented these events or Bretz's methods of inquiry. Bretz mapped the features of the Columbia Plateau. By reducing the scale of the landscape's features to map size (giant ripple marks, for example), he discerned patterns others had overlooked. His method proved adequate to the task, although decades passed before he convinced his colleagues that cataclysms on the Columbia had truly happened.

INTERPRETING THE LOCAL LANDSCAPE

Messing about with floods in a sandbox has a certain appeal among young children, but performing an experiment aligned with the inquiry jingle falls well short of learning to interpret landforms in the local environment. And interpreting landforms in the local environment would have been quite a useful and interesting outcome. The modern local landscape, clued by landforms, presents flooding and landslide hazards. Local businesses profit from copious deposits of sand and gravel.

The buried sediments of the Missoula Floods have even leaked carcinogenic radon gas into neighborhood homes.

Interpreting the landscape may progress across grade levels (fourth-graders need not be frightened by radon), and the stream-table-as-sandbox can teach a good deal about landforms sculpted by moving water. As an exploration and model deliberately linked to children's experience of their own landscape, stream table experiments have great value. Activity can be more than a venue for practicing stereotypical scientific processes. The stream table may augment but it cannot replace investigating the local landscape.

These fourth-graders lived amid a stunning and puzzling landscape indeed. Waterfalls predominate on one side of the nearby Columbia River Gorge. Across the river, massive landslides present corrugated profiles. Why were the waterfalls mostly on one side and the major landslides mostly on the other? Why were there exceptions to this pattern?

Instruction failed to challenge students to investigate local landscape phenomena using the stream table. There was too little time to couple stream table explorations to field study of local landforms and guide students on a virtual tour of the regional landscape. Instead, the fourth-graders proceeded to complete their lab reports synched to the one-size-fits-all-science scoring guide.

The landforms unit included both content and process objectives prepared for a curriculum that could be offered anywhere. The state tested for generalized knowledge of landforms, and the teacher assessed inquiry skill in the abstract. Neither assessment depended on knowledge of the local landscape or how scientists tackle problems of landscape change using methods tailored to field study.

A teacher ought to encourage children to engage in speculative discussion about the origins of local landforms and how to go about deciphering their history. After all, geologists have no freedom to conduct a controlled, single-manipulated-variable experiment to reproduce the patterns left in the wake of Ice Age flooding. The only stream table big enough is the earth itself. Bretz's inquiry was of a one-off, non-replicable experiment. He made mapping his method.

And so the study of landforms came to a close, and many opportunities to learn how geomorphologists actually study the changing landscape were missed. The potential for fascination with local geology was almost entirely overlooked. There was a great deal of attention given to three types of volcanoes: shield, composite, and cinder.

Ironically, all three were visible in nearby locations, if not from the schoolyard itself. A paper-and-pencil test assessed content knowledge; scoring a task assessed inquiry skill. Neither assessment captured the match between concept and method nor how geologic inquiry adapted to the demands of studying the local landscape, Bretz's story being the perfect exemplar. Generic, one-size-fits-all inquiry assessment and forced-choice/right-answer testing ruled the day.

The two-pronged approach—testing for content from the textbook and scoring inquiry completed as a work sample—socialized the children into the culture of school science. The culture conformed to what the administrator had shared with parents at Back to School Night: the dual-adoption (kits for process, text for content) approach. As an extra benefit, students learned the terms and definitions of landforms that they would no doubt relearn in eighth-grade earth science, satisfying the parent who spoke up to champion vocabulary.

Inquiry responds to the challenge presented by the nature of the phenomenon. The conception upon which all else depends integrates content and process. The official curriculum, the one wearing the mantle of accountability, ineluctably dilutes the local story. The view from the classroom window begs for an interpretive framework—a conception for assigning meaning to the stunning and varied contours of the land.

None of the NSF alphabets—not SCIS, not ESCP, not FOSS—and none of the NRC standards—neither the NSES nor the NGSS—seems poised to help fourth-graders appreciate the challenge of interpreting schoolyard and nearby landforms. An ambitious and well-informed teacher may do so, as many talented ones have throughout the region affected by the Bretz floods.

Yet the ambitious teacher cannot escape the demand to align instruction with state mandates and district adoptions. In favoring generalized content and universal processes, the standards-aligned curriculum denies stature and time to local story. The alphabets presume that scientific inquiry, as skills to master at the fourth-grade level, looks pretty much the same whether watching candles burn, whirlybirds spin, or stream tables drain. Those who know a particular subject well can sense the error in this presumption. It devalues topical knowledge and the interpretation of the local landscape.

ARCHAEOLOGICAL OUTLIER

Smelt runs—communities of eulachon, the oily "candlefish" of the journals of Lewis and Clark—nourished Native Americans in the Pacific Northwest for thousands of years. Their counterparts in the Finger Lakes region of central New York feasted on seasonal smelt runs as well. These local stories have the potential to enrich curricula across several fields of study. They are the stories of how people have inhabited the land over a very long time—often quite wisely. At the same time, they are understood in large measure through disciplined, archaeological inquiries.

What place does archaeology have in middle-school science teaching? Once upon a time, a dedicated group of archaeologists and science educators, inspired by graduate students who had shared their archaeological research experiences with local schoolchildren, made a case for putting zoo-archaeology at the center of middle-school science teaching, spread over three years. Their proposal failed, but lessons were learned.

In order to justify learning what archaeologists actually do to reconstruct village life from long-buried middens laden with smelt bones, the zoo-archaeology team learned that lessons must conform to generalized notions of hypothesis testing and data analysis in scientific practice. Archaeology's key concepts of "artifact" and "temporal horizon," to name only two, guide inquiry but do not pop up among the disciplinary core ideas that shape middle-school science. Moreover, archaeology suffers from the suspicion that the reconstruction of human society by means of archaeology is somehow a less scientific process than experimentation in chemistry, biology, or physics. This suspicion is the residue of faith in the scientific method.

The idea of linking a school to a local story—the archaeological record of how people exploited their surroundings compared with the present, for example—may hold promise and generate enthusiasm. The subdiscipline zoo-archaeology zeroes in on how people have interacted with animal communities in a place over time; it's a superb discipline for anchoring school learning to the study of place and fostering meaningful attachment to history and landscape among young citizens. But is it in the science standards?

To some extent, yes. Archaeology subsumes chemical analysis, radiometric dating, comparative anatomy, and ecology. Perhaps archaeology, even zoo-archaeology, works better than other contexts for inspiring students to learn core ideas of physical and life science. Becoming a means to that end, however, sacrifices the pursuit of what archaeologists do and why. Becoming a means to that end devalues archaeology's unique contributions to understanding the human condition, which are ends in themselves.

The standard question becomes, "Will this topic incorporate disciplinary core ideas and enhance understanding of crosscutting ideas and scientific practices?" All educators everywhere at all levels must ask this question repeatedly. The innovative educator, excited by the prospect of introducing today's students to the people who inhabited the same land long before them, seems caught between Scylla and Charybdis—crosscutting concepts and scientific practices.

In hastily linking to big ideas or the nature of science (NOS), in becoming the means to master core ideas in physical and life science, archaeology lessons dare not linger on how a particular set of concepts—site, artifact, midden, horizon—guides archaeological thinking. What archaeologists do when doing archaeology counts less than what they do that looks like traditional science.

As in geologic inquiry, ordering events in time is crucial to archaeological argument. As archaeology wrestles to tell the story of ancient people through time, it departs from experimental science stereotype. This departure makes its inclusion in school science suspect. Besides, given state testing and core idea progressions, school science is already overcrowded with what to cover. What the discipline must do on its own terms in order to solve its problems just does not rise to upper echelons of standards-driven reform, no matter how valuable disciplinary diversity and local story might be to student interest in science.

"Taphonomy" is a field of inquiry pertinent to the historical sciences such as archaeology, anthropology, and paleontology. Taphonomists study the conditions surrounding death. Often a zoo-archaeologist asks, "What does reconstructing the circumstances of death and decay reveal about habitats?" Understanding how creatures die helps to reconstruct the environment in which they lived. That's a good example of thinking archaeologically and paleontologically.

Bodies decompose and bones alter mineralogically. Nutrients recycle. Mapping these phenomena onto state and NGSS frameworks is possible, but it diverts instructional planning away from climbing the ladder of thinking archaeological thoughts—a ladder of expertise and purpose in its own right.

The ideas of growth and decay and life cycle play roles in taphonomy. Sleuthing the circumstances of death—as well as the preservation and alternation of remains—must dig deeply in order to uncover insights of environmental change through time. Zoo-archaeological practices, taphonomic and otherwise, do align with ideas of sustainability, biodiversity, biosphere, and earth processes in the standards.

Instead of worrying about such alignment, the representation of zoo-archaeological inquiry in school science ought first to attend to the questions it raises and the tools it uses to answer them. Among the questions are: What does the biology of animals reveal about human-animal relationships? Which birds, fish, reptiles, and mammals in the local environment have been important to societies in our region through time? How do scientists use radiocarbon dating and trace element/stable isotopes to reconstruct food pathways? What does reconstructing circumstances of death and decay reveal about habitats?

Proponents of zoo-archaeology in middle-school instruction believe in its potential for teaching students "to use the practices of science to participate in authentic local zoo-archaeological research." Ultimately, only teaching the generic practices of a singular science, not the proprietary aspects of a particular discipline, justifies zoo-archaeology in the middle-school science curriculum.

For sake of inclusion in school science, the proponents of zoo-archaeology described its practices in the inquiry language approved by the state: "Forming a Question or Hypothesis, Designing an Investigation, Collecting and Presenting Data, Analyzing and Interpreting Results."[2] The integrity of archaeological practice, the value of knowledge about local landscapes obtained through an archaeological lens, and the usefulness of specific ideas from taphonomy (and other fields) faded away in order to bring the curriculum into alignment with universal science—and the lyrics of the inquiry jingle.

Practices characteristic of geological inquiry, when organized to interpret place, emphasize temporally and geographically restricted solutions, not universal knowledge. Analysis of the case of the "Ghost Forests of Cascadia"[3] makes this point.

THE EARTHQUAKE MACHINE

Stick, slip—that is the basic mechanism of movement along a fault in the earth's crust, which results in an earthquake. Waves ripple forth as brittle fractures propagate, damaging buildings, changing coastlines, precipitating landslides, and sometimes generating tsunamis. Nearly all regions of the nation are subject to earthquake risk to some degree.

In the Pacific Northwest, the Cascadia Subduction Zone (CSZ) poses a dramatic threat. Conditions mirror the geology on the opposite side of the Pacific, where in 2011 the devastating Tōhoku earthquake and tsunami struck Japan. The last great subduction zone earthquake along the CSZ occurred on January 26, 1700. Rupturing extended north and south along a thousand-mile fault zone from northern California to Vancouver Island, and a tsunami caused damage even in Japanese harbors across the ocean.

Clearly, understanding stresses and strains in the earth's crust depends upon grasp of basic physical science: forces, waves, and states of matter. Yet haste in developing these ideas coupled with emphasis on stereotypical science (manipulating a single variable in a classical experiment) too easily obscures the value of methods of inquiry and concepts of events characteristic of the geosciences. If citizens are to take earthquake and tsunami hazard warnings to heart and support expenditures for mitigating risks through expensive engineering, they must trust the science behind the warnings. This same reasoning applies equally well to climate science.

Teaching the experimental method and developing basic physical science concepts have less potential to reach this goal than bearing witness to features of the landscape that have recorded past earthquake events. Present-day methods of sampling tidal flat and continental shelf sediments, dating forest declines across wide regions, and actually measuring deformation of the land in real time convinced geologists that indeed the Pacific Northwest had suffered cataclysmic subduction zone earthquakes and tsunamis repeatedly every few centuries for millennia. The event in 1700 was the most recent. Their simplest message: it could happen again tomorrow.

Many sectors of society share the responsibility of educating citizens and helping them prepare for natural disasters such as earthquakes. Science teachers have a special responsibility: establishing a basis for trusting the work of scientists. A generic sense of the scientific method does not suffice. The task of the science teacher is to make accessible the conceptual systems, investigative techniques, and forms of evidence upon which particular claims depend—to make the unobvious less so.

Good science teaching reveals how particular problems are solved in specific contexts. Citizen action to mitigate geohazard risks depends more upon citizen trust in geologic knowledge of the CSZ than upon modeling faults with sandpaper blocks to teach experimental design and introduce basic concepts of physical science.

Teachers must assess student learning in terms of state-mandated, generic goals while remaining faithful to practices and concepts deeply rooted in context. As a consequence, however, they often miss opportunities to make explicit the analogy between a lab model and real-world, geologic processes. In the end, expectations to teach from the perspective of universal method and cross-disciplinary concepts overshadow connecting the classroom to the local landscape and building trust in the methods of the geosciences.

Geohazards exist globally but cause harm locally, even if their effects are felt across a vast region. Schools play a role in educating local citizens to be prepared for geologic disaster. Widespread understanding of earthquake and tsunami hazards produces citizens motivated to take steps to mitigate risks. For example, coastal communities wisely locate schools and emergency services beyond the reach of tsunami inundation, mark evacuation routes, and construct vertical evacuation structures near beaches.

How do teachers typically teach about earthquakes? They often start with an attempt to demonstrate what happens when one occurs. The Earthquake Machine,[4] a simple sandpaper and wooden block demonstration, works well in this regard. It does not bring real earthquakes into the classroom but does introduce some of the physics needed to answer what happens when one occurs. However, the assessment materials for teachers accompanying this exercise short-circuit the complex task of linking the model to its real-world counterparts, such as the CSZ. In the end, even the earthquake machine succumbs to the temptation to teach universals of inquiry and develop basic concepts common to all subjects.

For several years seismicity workshops have introduced middle-school science teachers throughout the United States to the Earthquake Machine. This lesson, one among many from the professional organization IRIS (Incorporated Research Institutions for Seismicity), models the accumulation of elastic energy and its sudden release as "elastic rebound." The model consists of wooden blocks with sandpaper glued to one side, rubber bands, stiff paper, and a sandpaper track.

The modeler attaches a rubber band to a wooden block resting sandpaper side down on the sandpaper track and pulls gently with steady force. Friction holds the block in place until it suddenly releases and slides. Interestingly, the rubber band maintains some tension even as the block comes to a halt. Pulling with a steady force continues. The block slips again. The time between slips and the length of each slide varies.

From the stiff paper, often part of a manila file folder, the modeler fashions two rectangular frames that can wobble. When stuck to the pulled block, these frames bend back and forth when the block moves and stops, as do buildings on the earth. At this stage, the modeler attaches the second block to the first using another rubber band and repeats the first experiment, pulling with a gentle, steady force in a straight line. The first block, as before, slips suddenly, and the band between the two blocks stretches. The modeler continues to pull, and the first block slips again, increasing the tension, or stored elastic energy, in the rubber band connecting the two blocks. Suddenly the second block slips—the biggest jump yet. The two-block system accumulated the most energy, causing the dramatic elastic rebound of the second block.

The block, rubber band, and sandpaper system directly demonstrates the interaction of tension and friction forces and the exchange between potential and kinetic energy. The direct demonstration works perfectly well in a physical science class when teaching the concepts of inertia, force in the forms of friction and tension, and energy transfer. Students' grasp of these basic concepts may proceed rather straightforwardly. They can even experiment with the relevant variables by manipulating the sandpaper grit, the weight of the block, or the cross-sectional diameter of the rubber band.

At this level the Earthquake Machine treats seismic knowledge as a derivative of basic physics. That is certainly a truism, but achieving community understanding of earthquake risk requires more.

By analogy, the system models earthquakes. Thinking analogically substantially complicates the instructional goal. The task of the earth science teacher becomes to map the model onto elements of earthquake mechanics. Ideally, a teacher in the Pacific Northwest can establish correspondences between the elements and actions of the model with specific places, faults, and historical events in the local landscape. At that stage the Earthquake Machine becomes much more than applied physical science.

Mapping the demonstration onto earthquake mechanics in a Pacific Northwest subduction zone context presents confusion. Along the CSZ, the Juan de Fuca Plate collides with the North American one, compressing the coastal margin of North America as the ocean floor descends into the earth's mantle beneath the continent. The model includes no examples of collision and compression. Stretching is the primary mechanism. The rubber band stores the energy. Stretching the rubber band represents the crumpling, dragging, and arching of the margin of North America. Both store energy elastically, but the analogy stretches the mind: lengthening in the model, shortening in the real world. Releasing the energy causes rebound in both cases.

The descending and overriding plates are locked together along the subduction zone near this margin by friction. The earthquake machine strongly models this phenomenon: friction locks the block to the sandpaper surface below it. The analogy holds well.

Motions in the earthquake machine are all in the horizontal plane. When a subduction zone earthquake occurs in Cascadia, the "toe" of the continent flexes and releases stored elastic energy. Vertical displacements are characteristic. Coastal land extends seaward and subsides. The rapid toe-flex and seaward-slip during an earthquake displaces enormous volumes of water—a tsunami wave is born. Landward, as the toe subsides and the ocean floor tucks beneath the earth buckling the surface, elevations increase. The model captures none of these motions.

At depth, the locked plate zone transitions to a stick-slip zone, then to a continuously sliding one. By analogy, the two-block model roughly demonstrates how annual small earthquakes above the stick-slip zone in slow motion add to the strain of the locked zone, thus increasing the risk of megathrust events. The first block stands for the descending plate and its pull on the locked portion in its wake. The second block's sudden and dramatic jump models the enormous release of energy, its accumulation partially dispersed by the episodic tremors associated with movement along the semi-locked portion of the subduction zone. The sinking slab pulls, as does the rubber band. The locked zone is actually above the sinking plate.

Towns and highways known to residents sit above each of the three zones. Coastal landforms place geologic strata in unusual places. Tide channels uncover layers laden with remnants of upland spruce forests. Beach sands sit perched above the seashore. Cliffs of frozen lava that once filled marine channels stand in high relief.

What features in the landscape of Cascadia, cast in terms of plate tectonic theory, do the blocks and rubber bands actually stand for? When the blocks move, what changes in the local landscape do they represent? Where do the elastic rebounds modeled by wood, rubber band, and sandpaper happen in the real world? Where have they in the past? No one tells Pacific Northwest teachers in seismic workshops *not* to ask these questions of their students. But the assessment tools for the Earthquake Machine look first at using the lessons to teach concepts of force and energy. In the end, the local connections matters most.

Explanation by analogy—using one thing to account for something else entirely—is difficult. The model offers an explanation only after the concepts abstracted from the source are applied to the features of the target. "Elastic rebound" is the key abstraction, an energy storage-and-transfer event. Difficulty stems from using the demonstration to develop the meaning of this abstract idea while simultaneously using it to model an earthquake. For someone who already grasps the idea of accumulating strain (as in compressed rocks and stretched rubber bands), the analogy makes sense. For middle schoolers, just assembling this idea in their minds and being confronted with the obvious dissimilarities between the model and the true object of interest, the leap muddles the matter.

It takes a good teacher to coach successful leaping.

Over the course of twenty years, seismicity workshops have taught earth science teachers to pull stuck blocks and reinforce student grasp of basic physical concepts. What do the assessment materials accompanying the earthquake machine ask students to do? A typical workshop concludes with attention to assessment, first of "Energy Change" and then of "Inquiry."

Given a diagram of the block, sandpaper surface, and rubber-band system, students use labels and descriptive phrases to "identify the *form* of energy before and after each transfer" and "*where* the energy transfer happens." They do so for two energy changes. The assessment of content knowledge steers clear of analogy to crustal contexts or to rocks and plate motions. Correspondence to features of the local landscape and earthquake dates are not at issue. Teachers are accountable for student progress and achievement, and energy transfer no doubt holds a prominent position in state testing.

What of the assessment of "Inquiry" in the context of the earthquake machine? Other seismic lessons in IRIS workshops certainly center upon measuring the shaking, on depicting shear and pressure waves, and on picturing actual and theoretical deformations of the land. The earthquake machine lesson, nevertheless, does end with an inquiry scenario and rubric for scoring student accomplishment.

The inquiry scenario asks students to test a hypothesis composed of two controlled variables, a manipulated variable, and a responding one. The hypothesis appears as a question and a hypothesis; for example, "What happens to the amount of force required to move the block when the weight of the block is changed?" In hypothesis form, the question becomes: "The force to move the block varies directly with the block's weight." The sandpaper grit and the surface area of contact are kept constant, or "controlled" in the terms of the lesson.

An inquiry report states the hypothesis, lists materials, describes the procedure, displays the collected data, and draws a conclusion. The inquiry assessment rubric has two parts: an eleven-point scoring system for planning the investigation and a fifteen-point scoring system for writing the conclusion. The conclusion should answer the question, provide supporting data, and present an explanation. The plan ought to include steps of the procedure, definitions of the variables, and measurements to make.

In a nutshell, the earthquake machine in a workshop context perfectly portrays how teachers are taught to assess learning in science. Despite a characteristic geoscience topic, cross-disciplinary concepts are substituted for context-specific ones. The scientific method, as classical single-manipulated/controlled-variable design, is defined as the endpoint. The priority of these concerns circumvented the hard task of extending the demonstration by analogy to real-world earthquakes. Finally, the instructional materials missed the opportunity to adapt the model by analogy to the features known to residents of the local landscape.

Sciences that serve local interests bring the hazards of guiding instruction with the quest for scientific unity sharply to focus. The quest for unity influences process, NOS, and scientific practices objectives. They remain constant and general as subjects change. The quest also directs content instruction by directing attention to cross-disciplinary concepts. The geosciences, well suited to local interests, commonly experience pressure to do both.

In alerting students to the existence and causes of geohazards, the methods and concepts of plate tectonic theory, for example, deserve center stage along with techniques for interpreting landscapes (geomorphology). Plate theory is crucial to understanding earthquake risk. Properly interpreted, local landscapes reveal past catastrophes and present-day dynamism. The landscape is never static.

Schooling tends to value timeless, universal, and generalized understandings from the sciences over timely, local, and particular knowledge. In principle, the NGSS embrace investigations of dynamic, local landscapes and analyses of how cultures have inhabited the land through time only to the extent that these studies advance the paradigm of universality. If they get in the way—if they consume too much time on their own turf—their place is suspect.

Explicitly adapting the earthquake machine exercise to Cascadian audiences makes perfect sense. When done properly, interpreting how the stick-slip of the model corresponds to the CSZ develops specific disciplinary concepts (e.g., episodic tremors, subduction zones) *and* relevant cross-disciplinary ones (e.g., forces, friction). Attention to the practices of seismic research, such as interpreting the relative motion of the crust between coastal and inland measuring stations and profiling the earth at depth with acoustic signals, logically follows. The aim of such understanding is to mitigate the risks of living within the coastal communities of the CSZ.

FALSE RIGOR

In the earthquake machine example, generic aims, given the force of state policy defining scientific inquiry, undermined good teaching at the middle-school level. The study of landforms in elementary school prioritized stereotypical science and failed to help children interpret their own surroundings sculpted by Ice Age meltwater. Stereotypical experimentation dominated instruction in both situations, rather than a valid portrayal of how geoscientists interpret landscapes.

Monitoring school progress coupled to the desire to promote equity inexorably brings pressure to adopt standards that unify—standards that standardize what to teach. As a result, imaginative, diverse science teaching struggles for legitimacy. If students are not learning the same thing, how are we to compare them? Such supposedly self-evident logic dismisses questions of fundamental importance about the role of disciplinary diversity in school science. Ease of comparing student outcomes comes at a high price: misrepresenting the diversity of the sciences, eclipsing opportunities to ground science teaching in local landscapes, and silencing students' own questions.

The value of the knowledge serving the lives people live, not training in unity's dogma, ought to determine the answer to "What to teach?" In proper context, disciplinary tools—concepts and practices—serve purposeful ends inseparable from community interests, such as augmenting community resilience in the face of disaster. In a purposeful context, high-level abstractions may flourish alongside personal interest.

The unifying dimensions of the NGSS detour the ascent through disciplined knowledge, divorce inquiry practices from understandings of the phenomena they target, and devalue interests that are timely, local, particular, and personal. The proper context for learning science is a particular problem, not the big ideas common to all the sciences. If the diversity of such contexts creates assessment headaches, so be it.

The science standards contain the states' answers to "What to learn?" Challenging the science standards begins with asking a more fundamental question: "What to value?" Silencing the value question promotes consensus, which enables a political agenda. The silencing, as D. Bob Gowin's epigraph warns, also promotes false knowledge and encourages a debilitating rigor. The knowledge is false because it misrepresents the diversity of scientific practices and downplays how inquiry intertwines with understanding. Assessment debilitates learners by divorcing propositional knowledge from corresponding methods of inquiry. The science standards movement promises only a false rigor, a false rigor that enables assessment.

NOTES

1. John Eliot Allen, Marjorie Burns, and Scott Burns, *Cataclysms on the Columbia: The Great Missoula Floods,* 2nd ed. (Portland, OR: Ooligan Press, 2009).

2. Oregon Department of Education (ODE), "Official Scientific Inquiry Scoring Guide," http://www.ode.state.or.us/wma/teachlearn/testing/scoring/guides/2011-12/science_inquiry_gr6-8_eng.pdf (accessed January 28, 2015).

3. Charles R. Ault, Jr., "The Ghost Forests of Cascadia: How Valuing Geological Inquiry Puts Practice into Place," *Journal of Geoscience Education* 62 (2014): 158–65.

4. Incorporated Research Institutions for Seismicity (IRIS), "Modeling Elastic Rebound: The Earthquake Machine," *IRIS Classroom Lessons and Demonstrations* (New York: IRIS, 2015), http://www.iris.edu/hq/resource/redefining_an_earthquake_v12 (accessed January 28, 2015). Originally published as American Geophysical Union (AGU)/Federal Emergency Management Agency (FEMA), "Stick-Slip Movement," in *Seismic Sleuths: Earthquakes—A Teacher's Package for Grades 7-12* (Washington, DC: FEMA, 1995), https://www.fema.gov/media-library/assets/documents/15229 (accessed January 28, 2015). Complete activity adapted for *Teachers on the Leading Edge* (funded by the National Science Foundation EarthScope Project, 2010), aligned with standards, and including a scoring guide by Denise Thompson (Orting, WA: Orting High School, 2012) available at https://media.up.edu/Physics/TOLE/Plates&Earthquakes/LessonPlans/EarthquakeMachine.pdf (accessed January 28, 2015).

7

The Downward Spiral

Unifying abstractions divorce practices from content. Mastery of generic ideas such as modeling or designing investigations is extraordinarily difficult to assess, especially in ways that enable comparisons across groups. Reliable assessments of samples of student project work depend upon correlating the scores of at least two trained reviewers, which is an expensive process. Scoring by classroom teachers integrated with instruction works fine but does not produce data for making comparisons across populations.

As a result, standards-driven assessment bifurcates science education content and practices. Yet content stands alone for testing that serves the states' and nation's appetite for comparison data. High-stakes testing dwells upon propositional knowledge and skillful interpretation of tables and experimental designs in print format (digital or paper and pencil). Testing "hijacks" the standards, says Diane Ravitch, university professor and Undersecretary of Education in the Reagan administration.[1] Curriculum narrows in order to conform to "allowable content" by grade level, the progression of the core disciplinary ideas in the Next Generation Science Standards (NGSS) and the earth, life, and physical sciences content domains in the original National Science Education Standards (NSES).

The separation of inquiry from content for assessment purposes isolates disciplinary concepts from their role in generating appropriate methods of inquiry. Isolating content invites testing for propositional knowledge and reinforces rote learning. Often the safest strategy is to memorize. In an urban middle school serving poor, minority children, anxiety over test performance precipitated stifling instruction that failed to help students excel. The burden of rote learning simply discouraged them.

I LOVE YOUR QUESTIONS, BUT . . .

In a poor, urban neighborhood school serving a largely African American population, science teaching spiraled downward as the teacher stressed rote learning to improve test performance. The topic was plate tectonics. Attendance declined, grades collapsed, and discipline problems permeated the class. Text and recitation dominated classroom activity.[2]

Picture the teacher introducing the concept of seafloor spreading to a class of eighth-graders. It is week four of the geology portion of the earth science course. The teacher poses a problem: determine the age of oceanic crust at a given distance from a mid-ocean ridge. To complete this task students use dates based upon magnetic polarity reversals. The reversals "date stamp" lava as it erupts. As spreading proceeds, the freezing lava moves away from the ridge. As a result, dates for strips of rock on each side of the spreading ridge run parallel to each other and move farther apart with time.

The teacher interacts with the students. The students interact with each other and with the materials provided for doing the exercise. Eventually class ends, and the students are told to complete the question as homework and turn it in. An episode of instruction has run its course.

Over the course of two months, despite this promising exercise, the class spiraled downward in attendance and behavior as testing in science for Adequate Yearly Progress reporting neared. Simulations of plate tectonic phenomena in problem-solving contexts consumed large amounts of time. Memorization, decided the teacher, was a more reliable way to prepare for testing. There was pressure to raise test scores, especially on the earth science subtest. The text would be the guide.

The problem-solving class was atypical. Classes began with the teacher taking attendance followed by his checking for completion of homework. The school operated under a block schedule; thus, class met every other day for ninety minutes. On a typical day, for up to the first thirty minutes of class, the teacher moved from desk to desk with his grade book open, recording and rating homework completion for each student while the class looked up answers to chapter questions. This procedure guaranteed each student a moment of personal contact with the teacher on most days. A homework completion rate of 50 percent was the norm.

During the weeks prior to testing, attendance declined to 70 percent. The seafloor spreading exercise occurred during week four of the two-month unit. Following are flashbacks of earlier weeks before pushing on.

First Week

While the classroom teacher attended to a hallway altercation, a neighboring teacher stepped in to call the students to order. She asked authoritatively, "What page are you on?" Students recognized this signal as a familiar one to open their textbooks.

The text-driven instructional routine was familiar to both the neighboring teacher and the students.

When the classroom teacher returned, he admonished the class, as someone had placed a tack on another student's chair. Then he distributed a worksheet for students to complete on their own at their desks. The worksheet emphasized vocabulary, and there was virtually no hands-on laboratory investigation. They were at the beginning of their study of the rock cycle. As the teacher prepared to return previously graded papers, he reminded the class that answers were "in the book and on the diagram on the board."

Students were conspicuously unable to decipher the worksheet diagram. The classroom teacher carefully described the formation of each type of rock (igneous, sedimentary, and metamorphic). The teacher then asked, "Do you want extra time to work on these rock cycle questions, or do you want me to start the video?" The class voted to see the video on predator-prey relationships, not a topic germane to their study of the rock cycle. The video addressed biology concepts being reviewed for upcoming state-mandated science testing.

Second Week

Class the next week began with the students' attention focused on notes projected onto the screen at the front of the room. These notes summarized the three main types of rocks, for example, "Igneous rocks are rocks that are formed from cooled, molten lava." Students copied these definitions in silence.

After fifteen minutes, the teacher asked, "What are the three main groups of rocks?" Students read answers to these questions from their notes.

"Open to page ninety-seven. What is the rock cycle?" asked the teacher.

One student responded tautologically, "Rocks arranged in a cycle."

The teacher, following a common pattern, did not directly address this response. Instead, he elaborated upon this answer saying, "One kind of rock can turn into another."

Another student then read from the text, "Igneous rocks are classified according to their composition and texture."

The teacher stopped him and asked, "How are igneous rocks classified?"

The student answered and continued reading. Thirty minutes later the same student approached the teacher. He was unable to answer the homework question, "How does the texture relate to where the rock forms?" The word-for-word answer appeared in the text. The task was to find the passage.

The teacher took over the role of reading from the text and introduced a distinction between intrusive and extrusive igneous rocks. The terms were unfamiliar to the class, and there were no sample rocks to examine. To develop the meaning of "intrusive" and "extrusive," the teacher introduced an analogy between these terms and the personality types "introverts" and "extroverts." The analogy failed to direct attention to the texture patterns of rocks and associate them with processes of formation.

The teacher followed up with the question, "What does coarse-grained mean?"

A student called out, "It goes around like a race horse."

Another student called out, "Rough."

The teacher wrote "coarse" and "smooth" on the board and directed the class to "copy that down."

The teacher continued to pose questions about igneous rock formation. "Why do intrusive igneous rocks have larger crystals?" he asked.

Expressing himself perfectly, a young man called out, "Because they are hot and they can't bust; the smaller ones will bust."

Shortly afterward, conflict ensued between the teacher and another male student. The conflict escalated until the teacher ejected him from the class. The interaction consisted of a somewhat flippant verbal exchange, a reprimand, a warning, the student being seated in isolation, and his removal when the teacher said without anger, "Grab your stuff and come here." The teacher phoned the office, and the student departed. The class then worked in silence.

With ample time remaining in the period, the teacher announced, "Pages 100 and 104 are due today. You have half an hour. Be sure to write down the questions. I helped you with some. The first question is, 'How are igneous rocks classified?'"

The repetitiveness of questions and copying of notes from the board dominated classroom time daily.

Third Week

Several absences signaled the beginning of a trend. On average, from this day forward, one-quarter or more of the students failed to attend class.

The teacher spent some time engaging in one of his brief, yet not infrequent, sermons on proper classroom behavior. Later, when everyone was working quietly, he took time to praise them and commented that perhaps he had not done so often enough.

There were definitions of cell theory, protein, and element to copy from the board—concepts unrelated to the nomenclature of rocks. Both seventh- and eighth-graders copied these notes in order to prepare for the high-stakes, statewide science exam.

Once again, students read from the textbook. The teacher scolded, "You need to be paying attention."

At the end of class, a student earning an F remarked, "I'm just a kid, and I don't want to copy definitions all day." This student rarely attended class during the next six weeks.

Should educators take seriously the comments of thirteen-year-old, often truant F-student? Yes.

Nearly 50 percent of the students came to class without their textbooks; it was February, and daily use of the textbook had been the standard routine for some time. Study moved on from the rock cycle to plate tectonics.

Fourth Week

"Stop what you are doing for a second and look up here." The teacher had drawn a diagram of a mid-ocean ridge spreading center in order to help students solve a problem on the worksheet. "You guys remember what magma is? You have oceanic crust. It grows both ways. As it travels, the crust over here is going to be older. You need to find the mid-ocean ridge. We'll kind of go through the first one step by step." The diagram depicted the Atlantic Ocean floor. Lines traced rock of equal age along a north–south trend, labeled to give ages in millions of years.

This exercise required students to interpret a diagram, make a measurement, complete a conversion, carry out a calculation, and solve a problem: determine the age of oceanic crust presuming a uniform spreading rate from a mid-oceanic ridge. "The rate at which sections of ocean floor are moving away from the mid-Atlantic ridge can be calculated by dividing the distance traveled by the age," read the teacher from the directions at the bottom of the page.

"Someone pick one of those lines. I'm going to measure the distance in millimeters from the mid-Atlantic Ridge to that line . . ." (1 mm. = 65 km.). "Take 23 and multiply by 65. How far is it from the mid-ocean ridge to that line? It's right here. For answer number one, you can put down 1,495 km., but it's more important that you know how I got that." The teacher instructed the class to add five zeroes to express the answer in centimeters.

Class turned noisy as they began to work on this problem. Students asked for the answer; they were not prepared to carry out the measurement and calculation. The task was to measure in millimeters from ridge to dated magma several different isochrons away, multiply by 65 to convert to kilometers, convert kilometers to centimeters by adding five zeroes, then divide by age in years to give spreading rate in centimeters per year. Regrettably, student questions and performance suggested little or no comprehension of either the procedure or the concepts.

Students resisted the challenge and attempted to find ways of obtaining the right answer without working through the steps of the problem on their own. They found the demand of the exercise unfamiliar, and for the most part, exercised a veto over the expectation that they complete the task. Copying, memorizing, and looking up answers was the preferred routine.

Fifth Week

Eighteen out of thirty students were in attendance. Management problems vanished, and many students responded to questions. With dominant and often disruptive personalities missing, different students answered questions, and female students played a major role in class. At the start of class, the teacher offered praise: "I actually wanted to comment that this class recently is doing a lot better." "Doing better" correlated with declining attendance.

The teacher directed the students to copy definitions of plate boundaries from the screen: convergent, divergent, strike slip. "Who's not done copying this down?" he soon interjected.

Next he asked, "What is the continental drift theory?" For each question, students found an answer already written on the board. After students had answered several questions, the teacher asked, "If I turn the overhead projector off, how many of you can answer those questions?" He turned off the overhead projector and repeated the same questions as before, beginning with "What is the theory of continental drift?"

Someone called out, "When the continents are moving apart."

"There are three kinds of boundaries. What are they?" asked the teacher, who repeated the question and finally answered it himself. His next set of questions came from the textbook. However, 50 percent of the class neglected to bring books. "You're leaving me no choice. I am going to have to start writing detentions," he said.

Students wished to know if Alfred Wegener, the meteorologist credited with discovery of continental drift, became rich as a result of his theory. They expressed interest in the human side of his struggles to win acceptance for his theory among geologists who were skeptical of his meteorologist credentials. The teacher took little advantage of this opportunity to examine the subject from the point of human interest.

After a break, the teacher reviewed the questions once again, coaching them to succeed on the forthcoming chapter test. A female student repeatedly responded correctly.

"What is glossopteris?" asked the teacher.

"It's a seed?" came the tentative answer.

"No, it's a fossil of a seed," corrected the teacher.

Moments later the teacher explained that plates move parallel in a strike slip fault. "Why do plates move?" the teacher continued.

A male student responded, "Because lava moves in currents."

Another male student offered thoughtful responses, and the teacher prompted these by asking more "Why?" questions than in other lessons. In the last exchange of the day, the teacher asked, "Why would an earthquake occur at a mid-oceanic ridge?"

A student responded, "Because it is spreading. When they are moving, they got hooks, and the hooks connect."

The teacher elaborated upon this image of "hooks," explaining that ridges broke apart into many small strike slip faults. In this exchange the teacher successfully bridged student language and imagery and the formal discourse of science. Unfortunately the lesson never turned to the example of plate boundaries close to the school. It favored the universal and abstract curriculum of plate tectonic theory over consideration of local geologic problems.

Sixth Week

The teacher announced that the film *Big Daddy* would be shown as a reward for the good behavior the class showed the recent substitute. The movie would be shown on Friday, and there would be a test on chapter 3 on Tuesday.

In order to prepare for the test and reinforce student learning, the teacher had decided to review plate tectonics using a different text. He instructed the class to open the orange books in place of their standard texts. He wrote on the board at the front of the class:

> How well can you answer another book's questions in earth science? Write the answers only in complete sentences: Pg. 184 #1, Pg. 296 #1, Pg. 300 #1.
>
> Then look at the pictures on page 305. How many of the questions on 305 could you answer?
>
> Define as many terms as you can on page 184. Also define strike slip boundary, convergent boundary, and divergent boundary.
>
> Extra credit: Draw a detailed, colored picture of a convergent or a divergent plate boundary.

For the entire period the class worked on answering questions in a text different from the one normally used. The vocabulary and concepts were identical to the ones already studied. The teacher believed that finding the same concepts and definitions in slightly altered form and phrasing in this different text would reinforce their knowledge and make them better prepared to answer questions on the state-mandated, eighth-grade science exam.

Seventh Week

A new instructional cycle began on the topic of soil. Students received worksheet packets for the new chapter. The teacher introduced the weathering section of the text by saying that it included many of the answers needed to complete the packets. The text introduced students to the concepts of chemical and mechanical weathering.

One student sat without a book or packet. Another left the room without permission, flicking the lights on and off. A third had no book. The situation felt familiar. Then an unexpected exchange took place.

The teacher asked, "Are there any questions?"

A student wished to know, "What is the process that changes the chemical makeup of rocks?"

The teacher carefully answered. "There are a few different ones that can occur. Oxidation is one. Look at page 113. It talks about water and carbonic acid." Several moments later the teacher approached the student at his desk. "Did I answer that question for you? Chemical weathering can happen a number of different ways." A month ago the school had suspended this same student, on this day he was an exemplary student.

Eighth Week

Attendance declined to sixteen students. The period started with chasing and chaos. Several students arrived late; the teacher closed and locked the door to exclude tardy arrivals. Several lacked books, a chronic problem. One asked to return to his locker to get his book. The teacher refused to let him; the confrontation escalated, and the teacher sent the student to the office.

The teacher spoke: "Quiet down, please. I don't mind if you are going to be talking, but you need to be working. If you're not doing anything when I come around, I'm going to give you a warning because you're not following directions." Eventually all were silent.

The teacher directed attention to the notes he had projected on the screen reviewing mechanical weathering and exfoliation as caused by heating and cooling. He explained, "Exfoliation is when a rock splits in layers. The sun shines and heats up the rock on the outside but not the inside. In the night it cools, expands and contracts over and over, gets weak, and cracks off." He also shared a diagram of soil horizons from top to bottom: A topsoil, B subsoil, C parent rock, D bedrock in a mature soil profile, that is, "one that has had time to form."

The teacher announced a forthcoming unit test and led a review. He tried hard to help students score high by reviewing vocabulary and discussing sample items. The teacher did everything he believed he could to prepare this class to score well on the next exam. In addition he emphasized the need to use class time well, complete homework, and develop good study habits. He told the class that doing so would result in high grades. He provided them with notes that answered the test questions and permitted them to consult notes during the test. He all but completed all of the work for them. Nevertheless, very few did well; most earned disappointingly low grades.

TWO MONTHS OF ROTE LEARNING

During the course of the earth-science lessons, the classroom teacher expressed his conscientious intent to consider each student as an individual and not to think of any of them in terms of race, color, and ethnicity. At the same time, he was well aware of their cultural backgrounds, from Puerto Rican to Pacific Islander. He worked to have an amicable relationship with the class as a whole as well as a personal relationship with each student.

His classroom routine most often started with going desk to desk, grade book open, checking for homework completion and encouraging productive work habits. His insistence on good comportment and polite conduct was firm and consistent, with an explicit use of a warning before issuing any consequence for misbehavior or rude speech.

He saw the problem of achievement as one of completing in-class exercises and homework to earn points. He lamented how repetitious he had to be during lecture-discussions with definitions and scientific principles, such as types of plate boundaries. Typically, students read from the text, answered textbook questions, then worked on additional exercises that required them to find these answers in their books and write them down.

In place of time-consuming challenges to solve problems, teaching in this grade-eight classroom turned to locating and memorizing information found in textbooks. Questions had right answers that demanded very little reasoning. High-stakes, forced-choice, content testing in science paradoxically undermined good instruction in this eighth-grade class and reinforced the achievement gap.

The teacher issued warnings for failure to focus on work, excessive talking out of turn, disrupting others, leaving one's desk without permission, and disrespectful speech. Issue of more than two warnings within a week led to detention. In response to behavior that threatened harm such as putting tacks on seats, he threatened suspension.

The classroom teacher created a safe environment where students were not permitted to harass or disrespect each other, where consequences for misbehavior were well known and consistently applied, and where rewards were explicit. Students talked freely with him; both he and they made references to skin color casually, for example, when talking about ultraviolet radiation and sunscreen as part of a lesson on light. Often, students sought his approval; several openly wished to please him. Nevertheless, escalation of disciplinary responses to misbehavior often occurred.

Overwhelmingly, students responded to teacher-directed, text-centered activity. They focused their attention either on the teacher and his notes on the board or screen at the front of the room or upon worksheet packets that came with the textbook adoption that they completed either in class or as homework.

Recitation dominated. Frequent review, by means of chapter questions and homework assignments, characterized instructional style. Explicit rewards and reprimands preceded by warnings permeated teacher-student interactions. Often the teacher admonished the class as a whole about behavior and how behavior interfered with making progress. Popular rewards included entertaining films and breaks from the onerous routine of study.

Most students expended little effort; one-third failed to attend class regularly. The teacher's explaining, repeating, and restating lesson content increased in proportion to the steady downward spiral in student performance and attendance. Over the two months of rock cycle and plate tectonics study students almost never asked the teacher a question.

Rigor was misconceived as simple accountability for knowledge at a recall level and as disciplined work habits leading to rewards. Actions based on the classroom teacher's perspective, with its emphasis on rote learning, homework completion, and good behavior, did not succeed. Yet the cycle, intended to boost test scores, was safe and therefore popular.

CROSSCUTTING CONCEPTS AS A SOLUTION?

What do the NGSS have to say about this problem? The NGSS proclaim that integrating scientific practices with crosscutting concepts prompts students to reason in-depth and use "big ideas" to make connections across disciplines. The NGSS claim that opportunities to think in this way have eluded disadvantaged and underprivileged students. Interconnections are the missing ingredient. Crosscutting concepts, applicable across all the scientific disciplines, function to leverage sound grasp of each discipline.

The NGSS advocate setting performance expectations in terms of the standards. In performing a task, solving a problem, offering an explanation, or designing an investigation, students reveal their understanding of disciplinary core ideas and scientific practices. They do so in the context of the crosscutting concepts.

Such rhetoric clothes tired dogma in progressive-sounding language. The order of phrases once again does not matter. Students might just as well reveal their understanding of crosscutting concepts in the context of disciplinary core ideas and practices. There seems to be no relationship whatsoever between the NGSS's conception of promoting depth of understanding through contextualizing science lessons with crosscutting concepts and the week-by-week downward spiral in the earth science classroom.

An extraordinary disconnect exists between the scale of text-driven rote learning in this eighth-grade classroom and the grand scheme of using crosscutting concepts as a context for learning science. Even a modest effort to engage students in problem-solving frustrated the teacher. On one occasion students did work on solving a problem. The problem introduced seafloor spreading, and the well-defined context and a specific task served the purpose of stimulating thinking. However, the expectations of accountability on state tests subverted this effort.

For disadvantaged and underprivileged students, the reduction of learning in science to burdensome rote memory—in part a classroom management strategy—has a long history. This approach feels oppressive, and students rightly recoil.

Divorcing methods of inquiry from content knowledge enables the rote learning strategy. It lets content stand alone without relating it to methods of inquiry.

Proponents of the NGSS take as an article of faith that making connections across disciplines using a small set of big ideas will help students learn to reason in-depth. Making crosscutting concepts the context for learning science ignores students' own questions as effectively as it dismisses the reality of disciplinary diversity. With the contexts of personal experience and disciplinary coherence vanquished, meaningful learning has little chance to take root.

DEBILITATING RIGOR AS THE PROBLEM?

The text was the curriculum and the teacher drew from text-embedded and chapter review questions. The sterility of this approach invited expressions of oppositional, if not ambivalent, behavior, ranging from passive resistance to outright defiance: 50 percent of the class failing to bring textbooks to class and the high absentee rate.

Habits of rote learning subverted efforts to teach problem-solving. When presented with the task of calculating spreading rates, most tuned out, a few did the calculations, and several did not hesitate to copy answers from their peers. The teacher professed sympathy with the predicament of his students by saying, "I love your questions, but we have to get through this stuff." Tasks that confused students required too much time to sort through; simpler, rote ones moved coverage ahead.

The scoring and record keeping required to monitor the progress of students put at risk of failing and the accumulation of paper resulting from print-bound exercises subtracted time from planning for interesting classes. A great deal of class time was spent talking about work habits. Low levels of attention followed from uninteresting content; low levels of performance prompted the teacher to structure the class in increasingly less challenging ways.

Student attention declined and attendance dropped. The teacher applied pressure to students to discipline themselves to complete homework, study review questions, and answer chapter questions. A debilitating pattern of interaction reinforced itself: interesting content and complex, hands-on activities presented the teacher with overwhelming management problems; therefore, the teacher reduced the complexity and omitted challenging tasks.

Only a small number of students found the topics interesting or connected to personal experience. Those that in fact engaged in study—"good" students—became models for others. These others then mocked the good students or enjoyed labeling themselves and many of their peers as students who were "not good in science."

From plate tectonics to the rock cycle, good students primarily strove to complete, not comprehend, their work. Rigor misconceived as learning vocabulary propelled the curriculum into a downward spiral of note-taking and chapter question review. Students rarely answered why, made inferences, solved problems, or interpreted data. The teacher's lack of background knowledge of geoscience and his naïve conception of the nature of science compounded the problem.

The imperative of covering science content silenced student questions and reduced the idea of learning to remembering. Belief in this false idea of knowledge rewarded the complacent, who earned praise while disciplining themselves to memorize.

Unless broken, this cycle threatens to maintain marginalization. The cycle fails to craft experiences linking disciplinary practices to problems meaningful in students' personal lives. It fails to bridge personally meaningful expressions of ideas to the formal ones of science. Core progressions, scientific practices, and crosscutting concepts do not address this fundamental disconnect.

Whether by stressing universal abstractions or promoting rote memory of facts and vocabulary, school curricula eviscerate bodies of knowledge. Feeling the significance of a problem depends upon placing the problem within a purposeful context. The quest for unity divorces methods of inquiry from content knowledge. Divorced from each other, neither thrives.

The insight that bands of magnetism in oceanic rock track how the seafloor spreads was reduced to frustrating calculations made from measurements of a simple diagram. There were no connections to the local landscape resulting from this process—or of an active plate margin, which was quite literally the ground beneath the school.

Rote learning about plate boundaries and the rock cycle illustrates a classic problem for the education of poor minority youth in science: lowered expectations. Reform efforts that inadvertently place a premium on rote learning as a way of demonstrating literacy in science worsen the problem.

The authors of the NGSS do not disagree with such a conclusion. However, characterizing the disciplines in terms of a small set of practices common to them all and then embedding them in an equally small set of universal concepts does grave disservice to science education. The characterization depletes the store of disciplinary resources that make the sciences interesting and goes hand in hand with setting up content assessments that reinforce rote learning—and marginalize many.

NOTES

1. Diane Ravitch, "Hijacked! How the Standards Movement Turned into the Testing Movement," chapter 2 of *The Death and Life of the Great American School System* (Philadelphia: Basic Books, 2010).
2. Charles R. Ault, Jr., and Obed Norman, "Voices from the Classroom: African-American Students' Perceptions of their Science Classroom: Enhancing Performance through Understanding," Session 7.96, National Association for Research in Science Teaching (NARST) Annual Meeting, St. Louis, MO, 2001.

2

DIVERSITY'S STARTING POINTS

> A universal scientific method . . . common in discourse at all levels of science education . . . subverts young learners' understandings of both the practices and the content of the discipline.
>
> —Windschitl, Thompson, and Braaten[1]

The sciences engineer distinct concepts and methods adapted to the diverse challenges of inquiry. As phenomena vary, so do the sciences. Each discipline and subdiscipline responds in their own way to obstacles to inquiry and as a result develops characteristic ways of thinking and doing. Disciplines adopt different forms for representing data and expressing understandings. Investigative and rhetorical styles vary, as do explanatory aims. No alchemy exists to turn each and every discipline into physics gold.

Part 1 described the persistence of teaching science as process and the enduring appeal of the myth of the scientific method. Part 2 captures elements of diverse scientific practices that bring skepticism to this tradition with emphasis on examples from geoscience, the role of metaphor and analogy, and the distinction between experimental and historical styles of science.

NOTE

1. Mark Windschitl, Jessica Thompson, and Melissa Braaten, "Beyond the Scientific Method: Model-Based Inquiry as a New Paradigm of Preference for School Science Investigations," *Science Education* 92 (2008): 942, doi: 10.1002/sce.20259.

8

The Second Style

Disciplines adapt and disciplines diverge. Predictions differ in style from one field to another. At first glance, it would seem that all scientific disciplines are equally concerned with making reliable predictions, thus making reliable prediction a good criterion for judging the worth of an explanation from one field to another. However, the predictions of evolutionary biology differ from those of physical chemistry.

THE SINGULAR AND THE UNIVERSAL

Predicting reactions and predicting extinctions, or inferring chemical structures and inferring evolutionary ancestors, reflect different ways of dealing with uncertainty and different ways of representing reality. Chemists isolate reactions in laboratories; paleontologists search strata for the remains of life. Reactions have mechanisms; extinctions have histories.

The biologist typically predicts extinction for a species in probabilistic terms, using language such as, "With a high level of confidence, we believe that species A has a 50 percent chance of escaping extinction over the next century if the range and quality of its present habitat remain unchanged." There is uncertainty in the prediction as well as ambiguity in what "unchanged habitats" might mean. The evolutionary biologist may also make a prediction about the past, such as how long ago a lineage divided, for example, with one branch already extinct. Better evolutionary biology boosts "confidence" without promising certainty.

Classical physics adheres to different criteria. Its explanations represent interactions of objects mathematically using sets of variables bound to each other by fixed rules. One variable influences another deterministically. Equations of motion have universal application. In contrast, the colors in a geologic map and a Darwinian history represent singular chains of events.

Chemists calculate the rate, extent, and products of reactions with precision. A chemist's agenda is to predict with accuracy what new substances will form given initial physical conditions and knowledge of the chemical properties of the reactants. The chemist uses this knowledge to form substances with desired properties from substances with well-known yet quite different properties.

For example, at room temperature, sodium in its pure state explodes if mixed with water. Chlorine is a toxic gas at room temperature. They react in the presence of water to produce table salt and a great deal of heat (do not try this at home). A chemist typically casts a prediction in this language: "Sodium atoms will react with chlorine atoms in a one-to-one ratio to form table salt." Chemists and physicists both refer to particles. The chemist's particle is a space-filling object: the atomic elements bound into molecular compounds. Particles fill space, occupy positions, and have distinctive properties. Think ball and stick models. Often a physicist represents a particle as a dimensionless point, a mathematical abstraction. Particles move according to force, mass, and energy.

Neither representation contradicts the other, and both serve different explanatory ends. Biologists and geologists accept both models but practice sciences incompletely reducible to the study of particles. The emergence of living things from particle soup demands an appropriate understanding of cells, genes, and bodies. Physical science's fixed rules may apply universally, but they fail to satisfy as explanations for the existence of whales. The singular and contingent histories of complex objects, such as the earth, require historical approaches to explanation.

THE SECOND STYLE

The vision of unified method and common processes is a far cry from the opposite depiction of distinctly different scientific enterprises in loose alliance. The loose-alliance view distinguishes pragmatist or problem-solving agendas (e.g., climate change, evolutionary descent) from realist or truth-verifying ones (e.g., particle physics, chemical bonding). The loose-alliance view acknowledges a divide between experimental and historical methodologies even though both styles may complement each other within the same field. Historical and experimental styles merge when physics tackles cosmology.

Singular Events

Geology, the preeminent historical science lauded for its discovery of the geologic time scale, reconstructs earth history from the perspective of singular events as well as widely distributed processes. Geologists realize that this history is so deeply contingent upon past events that prediction of the future remains problematic if not impossible. Even when the future is deemed unpredictable, the past may remain explicable. Ordering events in time is prerequisite in accounting for the features of the complex present. Temporal order constrains possible causes. In the end, geologists do seek to link causes to their corresponding effects.

Stephen Jay Gould credited Charles Darwin with offering the sciences more than a conception of species evolving by means of natural selection. Darwin's narrative form of explanation, far from being merely a descriptive science, rose to the level of becoming science's "second style," the first being experimental method. Darwin adhered to the principle that all observation ought to be for or against some hypothesis, and he knew the importance of ruling out competing ones—by experimentation when feasible, by other means of comparison when not. His origins research, influenced by a conception of the earth's mutability over vast expanses of time, faced a primary obstacle: the frequent impossibility of experimenting on the scale of the phenomenon of interest to him. Other means were imperative and the second style descended from his triumphs.[1]

The Darwinian world is a contingent world where timing matters and temporal relationships are the key to solving evolutionary problems. Ancestors must be older than descendants. Similar thinking characterizes geological reasoning and figuring out cosmic history.

Darwin achieved early acclaim as a geologist. He returned from his *Beagle* voyage anxious to publish his ideas about the geology of South America. He explained South American topography from the Andes to the Pampas in terms of Charles Lyell's principles of uplift and subsidence. He extended these ideas to solve the puzzling variety of peaks, lagoons, and coral reefs among Pacific atolls.

Substituting Place for Time

As he climbed the Andes and sailed the Pacific, Darwin recognized something very special about the present: its landscapes featured examples at different stages of continuous geologic processes unfolding through time. Records of past events surviving into the present offered a sampling distribution of stages of geologic processes.

Volcano atolls (islands), volcanic atolls with fringing reefs, and islands encircling shallow lagoons: Darwin concluded that this seemingly anomalous collection of Pacific islands shared a common cause. They first rose, then sank, all the while with coral forming at a critical depth, building colony on top of colony. Active eruption lifted these structures. Once eruption ended, the island subsided. The scattered atolls of the Pacific visited by the *Beagle* sampled the stages of this ancient volcanic process. One island was an example of another's past; a different island was an example of its future. Islands substituted for each other in the hypothetical arrangement of stages in a continuous process through time.

The historical style generalizes his logic to become the principle "place substitutes for time." Continental margins, river valleys, volcanic landforms, plate subduction and rifting zones: for each of these phenomena, forms representative of different stages in their formation scatter the globe, the shards of a fractured puzzle whose pieces fit together in time. Confirming any arrangement of stages in time requires dating them independently of the stage-producing process. "Independent dating" is, in effect, a second principle of the historical style.

Interpretation and Scale

Deciphering the descent of an organism shares features in common with interpreting landscapes. In both, scale in time and place presents an obstacle. Uncovering the history of a landscape invokes modern analogues: looking elsewhere now for what was going on here in the past. Puzzling out a creature's Darwinian history utilizes homologies: comparisons of corresponding organ, limb, and genetic features from organism to organism, fossils included.

The historical style is an appropriate response to the demand for extrapolation across time and place on unobservable scales. This second style depends upon reasoning by analogy to render causal relationships plausible:

> Faced with the difficulties of modeling the geologic past because of problems of temporal and spatial scale and the singularity and complexity of geologic events, the geologist turns to other types of explanation. Arguments from analogy play a crucial role in the historical sciences; the assumption of analogy between past and present is what makes it possible to treat these subjects as sciences at all.[2]

Wrestling with scale permeates the second style. Confidence grows when solutions to problems on one scale reconcile with different problems on another scale. For example, widespread extinctions occurred at the end of the Cretaceous Period, approximately sixty-six million years ago, the date of dinosaur demise. Unusual amounts of the element iridium rest atop rock strata of this age at many locations around the globe. Iridium enrichment likely requires an extraterrestrial source. Are these phenomena anomalously associated or do they share a cause? Hypothesizing an impact crater holds appeal. An extraterrestrial object colliding with earth could have severely disrupted ecosystems globally.

Discovery of a crater of the right age and size beneath Chicxulub, Mexico, on the Yucatan Peninsula garnered support for the meteor impact or "Alvarez" hypothesis.[3] Was the crater of volcanic or meteoric origin? The meteor hypothesis reconciled a microscale pattern observed at the crater: "shocked" quartz (planar fractures). Only meteor impacts and nuclear tests are known to produce the pressures needed to create shocked quartz crystals.

The second style does not depend upon controlled experiments to rule out competing hypotheses. Instead, the second style relies upon a criterion of "integration across scales" and puts into practice the principle of "substituting place for time."[4] Furthermore, events thought to cause other events must cohere in time and place. Cosmic, evolutionary, archaeological, and geologic sciences practice various dialects of the second style. It provides the most telling example of diversity among the disciplines. Appreciating the importance of the historical style is the starting point for understanding the implications of the diversity of the sciences to science education.

THE ARCHETYPAL QUESTION

The quest to unify the sciences became an educational issue at the dawn of efforts to include training in the natural sciences as part of a university education. Whether to consider the sciences clearly united or significantly distinct held implications for organizing instruction as well as for the status of different disciplines. Were all the sciences equally scientific, or did some hold privileged positions?

The phrase "equally scientific" presupposes a unified view of science—a common standard by which to judge. However, conforming to a unified standard—for example, of experimental rigor—easily distorts how approaches to inquiry adapt to different challenges.

In the early nineteenth century, the positivist philosopher Auguste Comte reflected that physics had earned a position of privilege due to its success in discovering the lawful aspects of nature. Mathematical representation of these laws offered reliable prediction in many contexts. The laws of physics joined the explanations of earthly phenomena to those of the cosmos. Orbiting, for example, was a case of falling, both governed by Newton's universal law of gravitation.

Biology, thought Comte, had a ways to go. He speculated that the advanced, more lawful status of physics was due in part to there being greater complexity to biological phenomena and hence more difficulty in ascertaining the laws governing their interactions. Comte also concluded that biology was less experimental than physics. In place of experiment, he acknowledged, biologists invoked the comparative method pioneered by the French anatomist Georges Cuvier.[5]

Comte recognized that the comparative method, artfully developed and employed by Cuvier, was especially distinctive and dominant in biology. Comte attributed the ascendancy of the comparative method to its power to classify living and fossil creatures and unveil the natural order of relationships. He came to this conclusion before publication of *The Origin of Species* by Charles Darwin in 1859.

Cuvier's observations parted whales from the company of fish to give them membership in the society of mammals. His method gave birth to the discipline of paleontology and its successes in reconstructing extinct beasts from mere fragments of their skeletons.

The comparative method made a distinction between the structure and function of an organ. The method found ways to unmask natural affinities disguised by diverse adaptations. Using these very same homologies readily transferred to Darwin's evolutionary approach to solving problems of affinity and ancestry among living organisms. The focus on Cuvier's comparative method, flourishing in the present in evolutionary biology, anchored Comte's pioneering message that the sciences differ.

The Victorian zoologist Thomas Huxley (known to history as "Darwin's bulldog") dissented from Comte's implication that biology was somehow less experimental, less lawful, and hence less advanced than physics. Huxley rejected Comte's discipline-by-separate-discipline portraiture of the sciences and embraced instead the unity of science under the banner of experimental method, claiming biologists were as experimental as anyone else. For Huxley, controlled experiments yielded results interpreted by comparing them to predictions. Comparison was "the essence of every science" and was required to "discover a relation of cause and effect."[6]

Huxley objected to physics holding advanced status over biology (what he referred to as the "natural history sciences") and being somehow more scientifically rigorous. At the time, he was raising his voice in favor of incorporating the study of science into the general education of Great Britain's elites. The value of learning science, he reasoned, lay in its objective generation and justification of empirical claims, each such claim demonstrable in material terms and restricted to natural causes. Huxley expected science education to rout sclerotized church dogma, superstition, and belief in supernatural causes.

Debating the nature of scientific unity has been endemic to teaching science ever since Comte and Huxley devotees locked horns. From the perspective of scientists, unity promotes a sense of egalitarianism: "We're all equally good scientists at this table." From the perspective of teaching science, the unity view prevails in part because it immunizes disciplines, especially evolutionary biology, from viral attacks for being less scientific or less experimental.

The perspective of unity has proven very useful to those critical of Creation science. "There is no basis for separating the procedures and practices of evolutionary biology from those fundamental to all sciences," concluded noted philosopher of science Philip Kitcher in his popular book of a few decades ago, *Abusing Science: The Case Against Creationism.*[7]

Even as Kitcher defended evolutionary biology as inseparable from other sciences in terms of practices and procedures, he described how its characteristic methods of inquiry produced "Darwinian histories." Darwinian histories are explanations in narrative form that differ from mathematized physical laws.[8] Darwinian histories typify historical science.

Kitcher's two arguments—the importance of scientific unity and the existence of disciplinary distinctiveness—echo debate present at the origin of science education as a university subject, a debate inherited by K–12 educators. Kitcher's duality suggests a fundamental distinction between styles of science with important implications for school science.

OF CABBAGES AND HORSE TEETH

Despite unifying standards, diversity does manage to find its way into the classroom and do justice to both experimental and historical styles of science. Cabbage juice functioning as a pH indicator makes chemistry and the logic of experimental design approachable at a sixth-grade level. Using replicas of fossilized horse skulls, secondary students may learn to correlate skull morphology with habitat variation through time in the historical style of vertebrate paleontology.

Cabbage Juice

The experimental style is not the only style of science, but it does dominate school science. Done well and in the proper contexts, without being generalized as stereotypical scientific method, the experimental style rightfully claims a prominent place. A single-manipulated-variable design works well with sixth-graders investigating acidity. Changes in cabbage juice color, a response to pH, defines the responding variable in an observable manner.[9]

Picture pairs of student teachers joining several sixth-grade science classes to design and conduct investigations testing vinegar strength. The classroom is provisioned with a host of common, nontoxic household substances such as lemon juice, Sprite, clear shampoo, baking soda—and perhaps *very* dilute ammonia. In addition, there are several containers of white vinegar diluted with water to different degrees. Ample amounts of dilute cabbage juice are obtained by chopping, boiling, and straining purple cabbage.

The student teachers initially guide an exploration of changes in color when drops of the common substances are mixed in small tumblers of cabbage juice. Quickly students wonder whether they can turn the color back to purple again—or from pink to green and then to blue, or maybe even yellow.

Students observe changes in solution color with little difficulty and can easily sort substances into those that change purple cabbage juice to pink. Pink indicates acidic. Those that change cabbage juice to blue and even green are basic or alkaline.

Asking "Which vinegar is stronger?" or "Which antacid tablet works best?" challenges them to dig deeper. "Works best" and "stronger" are properties in need of definitions that can be used in making measurements. Calibrating the amount of vinegar needed to bring an alkaline solution back to a starting color poses a challenge to mathematical reasoning. A number of designs work reasonably well and lead to conclusions that take the form: "It took thirty drops of vinegar solution A to turn the alkaline solution back to the original cabbage purple, but we only need ten drops of vinegar solution B to make that happen. B is three times stronger."

Back on campus the student teachers tested antacid strength using the same approach that had fascinated sixth-graders—how to restore the cabbage juice to its original color. By extending the technique, they learned how to transform the investigation into a real-world problem. Complications arose because antacid tablets are mixtures of substances. Some tablets are buffered with salts of a weak acid, and some contain aspirin. Curious results led the teachers to inspect the ingredients listed on the antacid containers. Despite having gained some sense of how acids neutralize bases, a mystery remained—a good lesson for prospective teachers to take to heart.

Typically, reporting experimental results and focusing on the experimental process ends an episode of science teaching. However, methods and process are just part of the story. Difficulty in seeing what actually happens at the molecular level is a characteristic of chemical inquiry. This difficulty, an obstacle sixth-grade cabbage-juice chemistry cannot overcome, does merit attention in class discussion. Students should engage not only in defining and isolating variables but also in appreciating the obstacles to grasping the nature of chemical change.

Even work done as cabbage juice chemists depends upon a conception. The color changes students observed happened in response to mixing fixed ratios of acids and bases. Historically, discovery of fixed-ratio patterns among the volumes or weights of substances undergoing chemical changes proved indispensable to establishing the theory of atomicity. Investigating vinegar and antacid strength put the sixth-graders on a path toward understanding one of chemistry's most fundamental conceptions.

Cabbage juice chemistry teaches one way of honoring the diversity of the sciences. Learning the second style requires another subject entirely: paleontology does nicely. Both chemistry and paleontology taught well teach students to appreciate how *inquiry must adapt to the demands characteristic of the problem.*

Fossilhippuses

The Thomas Condon Paleontology Center has become the focal point of the John Day National Monument in Oregon. Across from the center rises Sheep Rock, a regional landmark capped with Columbia River basalt. Layers of volcanic ash reworked as lake beds and soil horizons, interspersed with local lava flows, span tens of millions of years of Paleogene history. Shades of blue, green, red, and tan hint at habitats long gone, ones much wetter than the arid present. Vertebrate remains permeate the ancient soils.

The Condon Center's dioramas tell the story of evolution in western North America from Eocene through Pleistocene time. Cheetahs once chased pronghorn here; today, no predator can keep up such speeds. Often standing in relief, hardened ignimbrites (fiery avalanche clouds of doom in their final resting place) and frozen lavas punctuate the soils and bottomlands of environments long past. Each of these volcanic deposits records immense volcanic activity and many can be dated radiometrically. They are the page numbers in the book of mammal history.

On a clear June morning, a group of high-school teachers in a field geology course rose and packed for a day in the field. Blue Basin was their destination. The monument's paleontologist had prepared a field mapping exercise. The class would work its way up a slope, planting small flags upon each sighting of vertebrate remains—small fragments of bone or tooth. Flagging had the potential of locating the source bed. In a previous summer, the paleontologist had unearthed a saber-toothed cat in a nearby location in this manner.

At the end of the exercise, the team decided to take one more hike along the upper trail. Midway through this hike, a flash of white against the bluish shale off the trail caught one of the teacher's attention. Bone? Yes, indeed—and not just a fragment. The bone, later exhibited by the Monument staff, turned out to be the fibula of an extinct, gracile-form rhinoceros. Gracile rhinos were much skinnier than the rhinos of today. Rhinos and their horse cousins roamed the West in many forms for tens of millions of years. Both groups sported an odd number of toes.

The Condon Center curates kits containing replicas of horse skull fossils that schools may borrow.[10] Needless to say, these kits are popular among teachers who have visited the John Day Monument, whether or not they have discovered any rhino fossils. Sixteen replicas of horse skulls with teeth from several extinct and one modern genera comprise the collection: *Hyracotherium*, *Orohippus*, *Miohippus*, *Archaeohippus*, *Hypohippus*, *Merychippus*, *Pliohippus*, *Equus*. That's a lot of hippuses linking the modern horse to its early ancestors.

In the first investigation, students attempt to determine horse size from skull length. The relationship between skull length and horse size has been remarkably stable: a multiple of 2.4 skull lengths does the trick. This pattern makes Cuvier's comparative method very useful. Entire skeletons are fossil rarities compared to skull fragments and teeth. Knowing relationships that allow for scaling a full-size creature from a small piece has served paleontology very well, thanks to Cuvier's eighteenth-century research.

Measuring length from the base of the neck vertebra to the tip of the front toe yields another critical insight. Whether or not the animal could reach the ground to graze (or a pool of water to drink) without bending its knees depended upon the sum of its head length and neck length. Being on bent knee in open territory increased the risk of predation. Predation pressure and habitat change likely combined to place selective pressure on horses, leading to the proportions observed in living zebras, for example.

Students do not have access to full skeletons. Instead, they are given the neck lengths and neck-to-toe measurements and prompted to hypothesize reasons for variation and the consequences of variation in different habitats. In this sense, they have added thinking in evolutionary terms to their experience of applying paleontological methods.

Next, working in teams of four, students turn their attention to the relative position of the orbit of the eye with respect to the teeth. The teacher asks them, "Which molar lies perpendicular beneath the front corner of the eye socket?" Three different molars occupy this position in the different skulls.

Teeth now take center stage, along with background knowledge that ridgeline patterns have increased in complexity over time. Sketching molar ridgelines provides data for a hypothetical ordering of skulls across fifty-five million years of horse evolution. Dating, even to determine relative age, must not be done in terms of tooth structure. Unless dating is independent of the ordering pattern, the temporal reasoning becomes circular.

With order in time established, students examine the skulls for other trends: skull length and relative position of teeth to eye sockets. Associating trends gives rise to speculations about diet, behavior, predation, and habitat. The geologic story of western North America is one of successively drier habitats and the expansion of rangeland. Individuals varied, and they survived to reproduce with different rates of success amid a transforming landscape. As a consequence, populations adapted. Grazers supplanted browsers. Extinctions and speciation among horses and rhinos proceeded for fifty-five million years, and then both groups departed the North American scene. Horses returned, via Spanish ships, in the sixteenth century.

The John Day Monument's horse fossil kit emphasizes variation, trends through time, and habitat change. It contains modules linking horse evolution to plate tectonics and examines the tooth morphology of carnivores and herbivores. A sibling kit explores plant evolution through another look at fossils. Leaves track changes in forest composition.

Horses did not evolve in a straight line of progressive change from multitoed browsing runts to modern Kentucky Derby winners. The fossil kit replicas sample only a portion of the bushy, branching equine tree of life. Different genera existed at the same time; the present is the anomaly, with only *Equus* trotting around (the horse, zebra, and ass genus). The kit organizes investigations of particular interest to students who live in the region; the place where their teacher found a rhino fossil is the place where they, too, might find a rhino fossil one day.

Most importantly, the horse fossil kit immerses students in paleontological methods and ways of thinking. It underscores the importance of ordering events in time to understanding the present. Exercises utilize the methods of comparative anatomy in an authentic context. Prior to the existence of the Next Generation Science Standards (NGSS), the kit content aligned very well with Oregon benchmarks in life and earth science, although mostly in a general sense (e.g., "understand interactions of living things . . . understand properties of the earth and how they change").

Generic, scientific inquiry objectives frame the kit: "Use interrelated processes to pose questions and investigate the physical world" and "Conduct procedures to collect, organize, and display scientific data," for example. Once again, these occupants of the driver's seat displace careful construction of objectives for thinking about thinking in context: how paleontologists go about solving their characteristic problems. Luckily, the generics are benign parasites. They do little or no harm. They simply draw time and attention away from the main story line.

The artifacts, though replicas, are genuine. The topic links to local experience, though it has widespread applicability. The science, though framed as generic, inescapably carries a paleontological fingerprint. Fossil science is historical, an exemplar of the second style. Too bad the top-down alignment mandate blocked a more careful development of a bottom-up analysis of how paleontological science gets done in response to the obstacles it faces. Benchmarks and standards, scientific practices and big ideas, are stuck-on afterthoughts that receive only perfunctory attention in the actual classroom. In the John Day Monument's horse fossil kits, fertile, organic ideas exemplifying Darwinian, second-style science get their due.

NOTES

1. Stephen Jay Gould, "Evolution and the Triumph of Homology, or Why History Matters," *American Scientist* 74 (1986): 60–69.

2. Robert Frodeman, "Geology as an Interpretive and Historical Science," *Geological Society of America (GSA) Bulletin* 107 (1995): 965. For a more complete discussion by Frodeman and a case analysis of geologic reasoning applied to the environmental costs of mining, see Robert Frodeman, *Geo-Logic: Breaking Ground between Philosophy and the Earth Sciences* (Albany, NY: State University of New York Press, 2003).

3. Carol E. Cleland, "Common Cause Explanation and the Search for a Smoking Gun," *Rethinking the Fabric of Geology: Geological Society of America Special Papers* 502 (2013): 1–9, doi:10.1130/2013.2502(01).

4. Charles R. Ault, Jr., "Criteria of Excellence for Geological Inquiry: The Necessity of Ambiguity," *Journal of Research in Science Teaching* 35 (1998):189–212.

5. Auguste Comte, *The Positive Philosophy of Auguste Comte, Volume II*, trans. Henrietta Martineau (Kitchener, ON: Batoche Books, 2000; originally published in 1830).

6. Thomas H. Huxley, "On the Educational Value of the Natural History Sciences," in T. H. Huxley, ed., *Science and Education* (New York: P. F. Collier and Sons, 1901): 48–49. Original work published 1854.

7. Philip Kitcher, *Abusing Science: The Case Against Creationism* (Cambridge, MA: MIT Press, 1982).

8. Philip Kitcher, *The Advancement of Science* (New York: Oxford University Press, 1993).

9. Jacqueline Barber, *Of Cabbages and Chemistry: GEMS (Great Explorations in Math and Science) Teacher's Guide* (Berkeley, CA: Lawrence Hall of Science, University of California, 1993).

10. John Day National Monument, "Horse Fossil Study Kit," *Traveling Trunk Fossil Study Kits* (Kimberly, OR: Thomas Condon Paleontology Center, 2014), http://www.nps.gov/joda/forteachers/materialsforloan.htm (accessed January 30, 2015).

9

Geological Reasoning

According to oral traditions, centuries ago indigenous peoples in the Pacific Northwest experienced a severe earthquake generated by the Cascadian Subduction Zone (CSZ).[1] Monstrous, inundating waves drowned shorelines villages. The quake was unknown to the current citizens of the United States until USGS geologist Brian Atwater began to puzzle over stands of dead cedar trees or "ghost forests" in coastal communities.

Tree ring dating placed their demise near the year 1700. He examined anomalous sand and rubble horizons beneath mudflats in tidal channels and wondered about even more deeply buried upland forest floors. Dating these phenomena at several sites along the Washington and Oregon coast converged on the year 1700. At about the same time, a tsunami was experienced along the shores of Japan.[2]

ATWATER'S GHOST FORESTS[3]

Along Cascadia's coast the death of cedars correlated with a peculiar pattern of sedimentary deposits: rubbly sand sheets interspersed with marsh deposits and mudflats. Perhaps local earthquakes liquefied sandy horizons beneath cedar forests, causing them to sink and drown. Curiously, at some locations fascinating geysers had ejected sand through overlying muds. In other places, low tides exposed root wads and spruce cones.

Farther inland, close inspection of tree rings revealed only disruptions in the growth of Sitka spruce and western red cedar. The timing of tree growth events chronicled sudden and synchronous destruction of coastal forests from Vancouver Island to northern California. The distribution of these phenomena raised a telling question: Was a subduction zone earthquake a regional risk?

As part of his investigative strategy, Atwater relied upon oral histories shared by Native Americans living along the Cascadian coast. Their history clearly included the experience of a great subduction zone earthquake and subsequent tsunami. The Coquille tribe, for example, recognized the potential for water to inundate the Coos Bay area.

Central to Atwater's strategy was to determine the sequence and synchrony of events. He needed to establish the timing order of geographically dispersed phenomena. In this approach, traces of events ordered and coincident in time created a set of anomalous associations. A convincing cause would be one rendering these anomalous associations plausible.[4] Atwater laid temporal traps to snare geologic cause. His conclusions proved convincing because his proposed causal mechanism resolved anomalous associations geologically as well as historically. His discovery of a Cascadian earthquake in 1700 turned a host of widely dispersed temporal coincidences into a coherent pattern of expected results.

The cedars died because the strain accumulated over centuries along the margin of the North American plate suddenly released. As the subduction zone ruptured, its undersea western edge lifted and thrust over the seafloor, setting a tsunami in motion. At the same time, its near-shore section subsided, in some areas by nearly two meters. The bottom fell out beneath the cedars. Ghostly sentinels now stand as witnesses to catastrophe. Harbingers of future tumult, they mark a significant horizon in time.

The timing of the most recent inundation by a sand sheet and the demise of the cedar groves coincided. On the other side of the Pacific, dutiful magistrates in Japan recorded the arrival of a series of tsunami waves at particular times and specific places along an extensive coast. Fortunately, these feudal Japanese documents, or "zassho," were carefully curated. They mapped the destruction of a tsunami with temporal precision from port to port and detailed the extent of damage in each harbor. However, Japanese documents failed to mention an earthquake. The tsunami was "an orphan."

Atwater's team analyzed the zassho records in order to infer the source of the orphan tsunami. They calculated wave direction, duration, propagation speed, and cresting height from damages to warehouses, homes, and rice paddies. These calculations pointed at a source; unwinding tsunami arrival times and intensity in Japan converged on the CSZ. Computer simulation of a CSZ earthquake verified their conclusion by demonstrating the capacity of the CSZ to create a tsunami capable of reaching Japan and causing the havoc chronicled by the samurai bureaucrats.

Lab physics and computer modeling made the zassho records of damage an expected outcome. The orphan had found its parent. At 9:00 a.m. on January 26, 1700, a one-hundred-by-one-thousand-kilometer section of the Cascadian Subduction Zone ruptured to produce a great earthquake (9.0 or higher) and tsunami.

Figuring out timing clinched Atwater's conclusion. Once convinced of the historical reality of one subduction zone earthquake in the Pacific Northwest, Atwater proceeded to extrapolate the past occurrence of others from analysis of buried horizons of marsh mud, tree roots, and sandy rubble. Once again, coordinating the temporal relationships among disparate phenomena distinguished his work. Temporal coordination of sand sheet horizons, cedar tree death, indigenous histories, and Japanese documents—abetted by computer modeling—proved productive.

The story told in *The Orphan Tsunami* stands as an exemplar of geological reasoning responding to the challenge of scale. Atwater made use of analogy in comparing one geographic region's history to another's. He matched historical dates from Japanese calendars to ages from tree rings and ran computer models to re-create the timing and extent of tsunamis in Japan originating from the Pacific Northwest. Analogs from subduction zone earthquakes in Chile in 1960 and Alaska in 1964 provided independent evidence for interpreting phenomena in Cascadia. Atwater found the evidence sufficient to rule out competing hypotheses such as inundations from local storm surges depositing sandy rubble and drowning cedar trees. Land over a broad region had indeed plummeted downward, crept upward, and done so again many times.

Investigators examining undersea landslides on the continental margin of Cascadia have independently confirmed the recurrence of great subduction zone earthquakes in the region. On average, a great earthquake spanning from Vancouver Island to Northern California occurs about every five hundred years. Given the time since the last great rupture and the irregular distribution of such events over the last ten thousand years, there is a 10 percent chance of a great subduction zone earthquake and accompanying tsunami happening within the next half century.[5] Citizens of the Pacific Northwest must confront in their own neighborhoods the very real dangers of a great subduction zone earthquake and, in coastal areas, an accompanying tsunami.

Thanks to Atwater's achievements, teachers have the opportunity to tell a local story of how geoscience gets done. Atwater's inquiry exemplifies the mix and match of specific research practices adequate to solving a particular geologic puzzle. The seismic story of one place on earth inevitably has analogues around the globe. Through the lens of plate tectonics, even ver local indigenous oral traditions have global implications.

REVISITING THE SUBSTITUTION OF PLACE FOR TIME

Analysis of geological inquiry illustrates the diversity essential to scientific enterprises. Methods of inquiry and conceptualization of phenomena are mutually dependent. They begin to shape each other as attention focuses on particular events and objects of interest—for example, the geologic processes that sculpt the earth.

The present captures patterns of geologic processes that cannot be directly observed on the human time scale. The vast time scale for these processes presents a distinctive challenge to geological inquiry. In response to this challenge, places often substitute for time. Scattered about in seemingly haphazard patterns, they record sequences of changes in landscapes that unfold slowly or happen cataclysmically. Geologists must confront a higgledy-piggledy data set, as opposed to records obtained systematically in laboratory settings. They understand the present as a sampling distribution of the results of geologic processes operating over long spans of time.

In adapting to the challenge of time scale by reconciling distributions of dead cedars, sand sheets, Native American oral history, Japanese zassho, and computer models, Atwater's inquiry displays many qualities that distinguish geoscience. His story of an orphan tsunami respects the diversity of scientific enterprises and how geological inquiry struggles to determine synchrony and correlation in order to arrive at a causal explanation.

The previous chapter credited Charles Darwin's Pacific atoll theorizing for demonstrating the power of the historical style of science. He pioneered how to substitute place for time in arranging present-day geologic phenomena as stages in a geologic process. Substituting place for time, a telling principle of geologic reasoning, is a method of inquiry responsive to the challenge of time's vastness. Characterizing geologic patterns and processes observed in present time as a sampling distribution of past and future ones enables extrapolations of geologic processes. One place may stand as an example of a past stage, another as an even more ancient pattern.

Substituting place for time does entail a risk of circular reasoning. Hypothesizing the order of events in time according to some causal principle and putting events in a causal sequence of stages must have independence.

For example, Cascade stratovolcanoes are presumed to be symmetrical and conical in a youthful stage, then broken and craggy in a later stage, the consequence of eruptive and erosive processes. If this arrangement in stages were exploited to determine relative ages, a craggy volcano would be labeled "old." However, using the stages to infer order in time is circular. The craggy volcano might be found to be youngest of all and a symmetrical one by far the oldest within a continuous range. That's what turned out to be the case in Oregon.

Radiometric dating gives the proper ages of the peaks of the Central Cascades. Black Butte, near Bend, Oregon, is a conically symmetric volcano. In contradiction to the expected sequence of stages in a Cascade volcano, the symmetrical Black Butte proved older than its craggy siblings. How did it escape the fate common to the High Cascade volcanoes? Was it composed of some anomalously resistant rock? No. Due to its location to the west of the High Cascades in a rain shadow, Black Butte escaped the heavy snowfall responsible for Pleistocene glaciations that carved and fractured nearby peaks. Just being dry confused the interpretation of order in time.

Mount St. Helens, on the other hand, is a very young and now quite craggy volcano because one of its slopes failed catastrophically in 1980. Getting these eruptive stages in proper sequences and recognizing the exceptions depends upon determining order in time independently. Understanding the anomalies—Mt. St. Helens's missing flank, for example—creates the opportunity to revisit analogous anomalies elsewhere around the Pacific Rim including just across the river. Mount Hood once lost a flank, too.

The challenge is to arrange places in a convincing order representative of a temporal process. Places, broadly conceived, include present-day landforms as well as buried, ancient landscapes. Atwater, for example, exhumed marsh-scapes from millennia past. Place substituting for time has pragmatic value but is not a geologic "law." It's a principle for guiding thinking and doing that overcomes the obstacle of scale when applied in concert with other approaches to interpreting the earth.

TEMPORAL REASONING

Not long ago the Denver public school system instituted a "Contingent Pedagogies" project to improve sixth-graders' knowledge of earth science. The project aimed squarely at representing the "practices [that] reflect the diversity of what scientists do." "Diversity" meant adding the practices of "developing and using models" and "engaging in arguments using evidence" to the time-honored "planning and carrying out investigations."[6] Subsumed under these abstractions, the intent of teaching earth science in Contingent Pedagogies was to demonstrate what unites the sciences. Scientists design investigations *and* use models.

Contingent Pedagogy weakly recognizes practices contingent upon a discipline's distinctive challenges. The common habits of mind matter most. Methods of inquiry invented to advance temporal reasoning—determining sequence and synchrony—may eventually find their way into Denver's sixth-grade classrooms, but not as the most valued aim.

Denver sixth-graders might find Cascadia's ghost forests quite fascinating. In their region, however, the puzzle of why the mid-American continent stands in such high relief merits scrutiny first. The history of collision between the North American plate and its Pacific plate neighbors figures in the gradual rise of the Colorado Plateau. Plate collision also caused the sudden demise of Cascadia's Sitka spruce and red cedars. Both puzzles call upon putting the practices of geoscience to work; both provide purposeful contexts for learning disciplinary ideas. Both lead to an exploration of the mechanisms of plate tectonics linked to the practices for figuring out regional geology.

Casting these puzzles simply as examples of the diversity of what scientists do mischaracterizes the practices valued in geologic inquiry. It teaches that all sciences do the same diverse things, but in different ways. The key point persists: scientists, even geologists, develop models and design investigations.

The weak acknowledgement of diversity unwittingly discounts expertise. Modeling, explaining, investigating, arguing from evidence, posing questions: this list admirably calls upon students to think. But it fails to capture what to learn in order to think distinctively geologic thoughts and to solve problems characteristic of geological—or any other—carefully contextualized inquiry. The list pertains equally well to any science, and that, unfortunately, is its drawback.

THE LESSON OF THE GHOST FORESTS

All geologic insight attaches to a sense of place in time; the risk of another great subduction zone earthquake in Cascadia is palpable. Residents find themselves ensconced in a time horizon bedeviled by this risk yet enriched by the majesty of their surroundings—Cascade peaks, rocky headlands, coastal dunes, and Sitka swamps. The risk and the majesty, cobbled together through time's vastness, share common cause in the mechanisms of plate margin tectonics. Nature's artistry has a price tag.

The Atwater story underscores the instrumental value of geologic knowledge to people who live in a particular place. Equally importantly, the story reveals profound shortcomings in the depiction of the sciences as a universal enterprise. Temporal reasoning, a hallmark of geological inquiry, distinguished Atwater's research.

Understanding scale is crucial to geology as a way of knowing. Geologic problems cross many scales, and at each scale solutions may differ. Yet these solutions must cohere—the explanation patterns of crystal formation must not contradict the account of regional crustal movements, for example, when interpreting a particular place on earth.

The present contains remnants of the past distributed as confusing puzzles. Confidence in geologic solutions increases when independent lines of evidence converge on the same explanation and solutions to problems on different scales cohere.

Temporal-spatial dialogue permeates geologic reasoning. Geologists depict challenging temporal data in dramatic visual forms. Geologic reasoning profits from rendering data into stratigraphic columns, correlation charts, block drawings, and cross-sectional diagrams. Its preeminent product, the geologic map, codes time with color. Geologic maps feature the contours of time. They display the horizons of significant changes in geologic processes.

In resolving geological puzzles, the strategy of substituting place for time often generalizes to become an appeal to modern analogs. Analogs are other places geologically similar to the place of current interest. Modern analogs for rupture along the Cascadian Subduction Zone are frightening: Banda Aceh, Indonesia, 2004, and Tōhoku, Japan, 2011. Chilean and Alaskan analogs are equally worrisome. Any timely, local story has global tentacles; the ghosts of the Cascade forests haunt many places.

The events and objects of interest to geoscientists—historical phenomena that reside in deep time and complex systems on multiple scales—present distinctive challenges to thought and inquiry. In order to interpret the earth, both the methods of inquiry and the conceptions of phenomena must match the nature of these challenges. Disciplines are disciplined by this match. The match bolts thinking to doing—conceiving to investigating—with clear purpose in mind. The geoscience match between temporal reasoning and dating techniques is another good starting point for appreciating disciplinary diversity.

NOTES

1. Ruth S. Ludwin, Robert Dennis, Deborah Carver, Alan D. McMillan, Robert Losey, John Clague, . . ., and Karen James, "Dating the 1700 Cascadia Earthquake: Great Coastal Earthquakes in Native Stories," *Seismological Research Letters* 762 (2005): 140–48.

2. Brian F. Atwater, Musumi-Rokkaku Satoko, Satake Kenje, Tsuji Yoshinobu, Ueda Kazue, and David K. Yamaguchi, *The Orphan Tsunami of 1700*, Professional Paper 1707 of the United States Geological Survey (Seattle: University of Washington Press, 2005).

3. This chapter consists in part of excerpts from Charles R. Ault, Jr., "The Ghost Forests of Cascadia: How Valuing Geological Inquiry Puts Practice Into Place," *Journal of Geoscience Education* 8 (2014): 158–65.

4. Carol E. Cleland, "Prediction and Explanation in Historical Natural Science," *British Journal for the Philosophy of Science* 62 (2011): 551–82.

5. Chris Goldfinger, C. Hans Nelson, Anne E. Morey, Joel E. Johnson, Jason R. Patton, Eugene Karabanov, Julia Gutiérrez-Pastor, Andrew T. Eriksson, Eulàlia Gràcia, Gita Dunhill, Randolph J. Enkin, Audrey Dallimore, and Tracy Vallier, *Turbidite Event History—Methods and Implications for Holocene Paleoseismicity of the Cascadia Subduction Zone*, Professional Paper 1661-F of the United States Geological Survey (Reston, VA, 2012), http://pubs.usgs.gov/pp/pp1661f/ (accessed February 9, 2015).

6. William R. Penuel, and Angela H. DeBarger, "Contingent Pedagogies," in *Education Views* (Boulder, CO: School of Education, University of Colorado, Spring, 2012): 5. The Contingent Pedagogies (NSF grant no. DRL-0822314) website is http://contingentpedagogies.org/.

10

Metaphor and Analogy

Pulitzer Prize winner Richard Hofstader considers analogy "the lifeblood of human thinking" and "the very blue that fills the whole sky of cognition."[1] In an impressive string of metaphors, Hofstader explains that thinking transfers "tightly packed mental chunks" that reside in a "dormant" state in long-term memory to short-term memory for "unpacking" in a series of "leaps."

Metaphors and analogies do the heavy lifting of mental chunking, transfer, and unpacking. One very famous leap in thinking aided and abetted by a metaphor comes to mind: from the image of a spiral staircase to the double helix structure of the DNA molecule.

The operation of mind depends deeply upon analogy and its less analytical, yet equally powerful cousin, metaphor. Metaphor and analogy are lichen-like symbionts: metaphor, the algal partner, providing the nourishment for new conceptual growth, and analogy, the fungal one, the hospitable substrate and structural ladder of thought. Metaphor defines analogy; analogy extends metaphor.

DISCIPLINARY DIVIDES

Metaphors guide thinking, dividing thought into pathways leading to distinct ideas. Some sciences isolate components of the natural world, reducing complexity and examining the most salient causal factors. Detractors of the reductionist approach may find this to be "mechanistic." Supporters value the universal applications of the fruits of disciplines such as "particle" physics and "string" theory.

The watersheds of the sciences descend from their divides. Some sciences accept the emergence of complex phenomena with patterns of interaction and causation irreducible to their antecedent components: the "tree" of life, for instance. Its phenomena of interest are "emergent."

Some sciences straddle the reductionist and holistic realms by experimenting in virtual worlds created by computational models: the climate scientists come to mind. The object they study is very much a pure analogy.

Little quarrel exists over describing the sciences as empirical and recognizing the criterion of replicable results. In addition, probabilistic and deterministic modes of thinking, represented mathematically, crisscross from one field to another. All investigators attempt to constrain uncertainty in their conclusions about causation. Yet such commonalities mask the interesting and diverse ways that disciplinary thought ensnares phenomena, inevitably by invoking metaphor and analogy. The imagery of their unity should not mask the exploration of the diversity of the sciences. Metaphors capture distinctions essential to understanding what scientists do and why.

Often in the sciences metaphors function to jump-start thinking, form new concepts, and communicate ideas. A proper analogy may accelerate a discipline's agenda and an implicit, unacknowledged one may hinder it. Disciplinary agendas differ. Metaphors often become apparent in how disciplines represent reality; paying attention to metaphors from one science to another unveils these differences. In brief, different agendas and different representations track different metaphors, and the metaphors are essential to creating meaning.

For example, physicists search for universals of causation, for the most fundamental relationships among matter, energy, space, and time. Physicists represent nature's fundamentals with mathematical objects and express relationships among them with graphs and equations. This rigor comes with a cognitive price tag: their work assumes a tight analogy between mathematical representation and reality.

Of course, mathematics, as a language of science, permeates many disciplines, but in physics the goal of mathematical expression—to capture universal features of reality—stands preeminent. Mathematical representation in quantum theory makes the visualization of phenomena of interest virtually impossible and invites fanciful nomenclature: quarks (which make up protons and neutrons) come in six "flavors," one of which is "charm" and another, "strange." The vivid names aid in keeping track of quark properties.

In quantum theory, photons are simultaneously analogous to waves and particles, making *Star Trek*'s photon torpedo a kind of ballistic weapon that surfs to its target. Quantum theory marks the extreme position on the spectrum of the uncertainty principle: reality depends on how you look at it.

Chemists aim to predict with computational precision how material substances, under a given set of conditions, interact and change. The chemist's tools of thought function to overcome the obstacle of invisibility at the molecular level. Visualization of the phenomena of interest enhances chemical thought. For example, the opposite ends of carbon "rings," the key player in organic chemistry, may flex to change the molecule back and forth between "lawn chair" (one end up, the other down) and "hammock" (or "boat") forms (both ends "up"). The molecule "relaxes" in either position.

Ball-and-stick models, dot diagrams, and lines whose length and angles represent atomic order permeate chemical representations. Picturing structures is crucial, and often the geometry of particle configurations holds clues to reactivity. They enable thinking about rotations, symmetries, mirror images, and, of course, substitution, elimination, and synthesis reactions. The balls and sticks drop off and add on.

Biologists value ideas that simultaneously account for the diversity and similarity of living forms. The imagery of both continuity and novelty predominates in biological thought. DNA sequences that persist through time define lineages. Alterations of ancient genetic codes turn scales into feathers and transform dinosaurs into birds. As presciently drawn by Darwin, branching diagrams, metaphorical trees and corals, have come to represent pathways of descent through time.

Observations of shared and unshared sequences of DNA across groups of microorganisms challenge the ubiquity of the ever-branching, never-fusing image. Now anastomosing or reconnecting, network diagrams suggest fusion events among lineages, especially during the earliest epoch of prokaryote, eucaryote, and protist life on earth. Parasites have taken up residence inside cells as omnipresent cellular machinery (that's three metaphors in the definition of "mitochondria").

Novelty may result from swapping genes across lineages. However, purely branching imagery sets up a barrier to this insight, failing to anticipate how genetic engineering would make tobacco leaves glow with jellyfish light.

Cosmologists, in extrapolating the guiding imagery and theory of physics to the universe as a whole, confront conundrums and contradictions. Maybe the universe is mechanical, maybe it's organic, or perhaps it's just computational. Perhaps analogy to natural selection may offer an answer to the fundamental question, "Why is our universe the way we find it to be, hospitable to humans?" Cosmologists have begun to wonder, similarly to Rudyard Kipling's musing about how the leopard got its spots, "How did the physical constants of the universe get their values?"

Thinking evolutionary thoughts about the universe as a whole—from big bang to big crunch, or perhaps its heat death—tends to provoke departure from the leading paradigms of physics and entertainment of more Darwinian imagery: the descendants of black holes varying by chance at the dawn of innumerable big bangs generation after generation.[2]

Geologists reconcile landscape patterns with processes operating on vast scales in time and place. The imagery of a rigid globe impeded progress in geologic theory, and until relatively recently, discounted ideas of continental drift. Continents, like bumper cars in an amusement park, collide, crumple, and clump, then disperse again. Historically, geologists have debated whether the earth overall remains in a steady state. Stephen Jay Gould encapsulated this debate as the contrast between the image of time's arrow and time's cycle. Has the earth simply run through cycles again and again over vast eons, or has its trajectory through time been an irreversible, ever-changing one?[3]

The imagery attached to conceptions of time varies by discipline, reflecting whether theoretical constructions cast time as real or illusory. Ferreting out relationships in time may confront psychological barriers, or time itself may function as a referee among competing hypotheses. No one can feel the difference between ten million years and ten billion years. Forty thousand years, a vanishingly small sum on the cosmic calendar, sounds like a long time. Ordering events in time—fish fins before reptile limbs—anchors plausible evolutionary arguments, providing the dating is trustworthy. Time is both a barrier to easy insight and a criterion of good explanation.

The Next Generation State Standards (NGSS) are silent on metaphorical underpinnings of theory, the importance of imagery to guiding thinking, and the aesthetic appreciation of natural phenomena. There is artful perception and expression in any science. Standards neglect to mention the poetic fierceness of the scientific imagination. States assess cognitive outcomes, not expressive ones. The literary touch—the story that tunes the parameters of meaning—codifies poorly. Great teachers acknowledge the metaphorical components, the guiding imagery, and the aesthetic dimensions of the subjects they teach. At least art dwells on the walls of science classrooms.

METAPHORICAL TIME

"Time" in the sciences finds expression in several metaphors. Sometimes geologists refer to a "cosmic calendar" in order to grasp time's vastness. "Timelines" express time as length, which time is not. Many disciplines find merit in the "arrow" of time. "Clock" time crops up as a variable in mathematical expressions of time. Choosing a metaphor carries consequences for understanding within a discipline, and understandings of time inevitably come wrapped in metaphor and analogy.

Arrows and Clocks

The arrow of time, a metaphorical expression of the second law of thermodynamics, takes as its target the clockwork universe. Disorder increases; clocks run down. The direction of time's arrow is irreversible.

The clockwork metaphor, on the other hand, suggests symmetrical time: the laws of physics winding and unwinding equally well whether run forward or backward in time. The clock's unwinding sequence is logically present in its initial conditions. The clock is a metaphor for determinism—once the parts are configured, all the outcomes are predictable—or deducible.

Theoretical physicist and science philosopher Lee Smolin finds many arrows in time's quiver. Time's thermodynamic arrow flies with increasing local disorder. The cosmological or expanding universe traces an arrow of time. There's no reversing the arrow of biological growth and decay. Time's arrow orders experience and memory.

Smolin accepts the idea of an electromagnetic (or light travel) and the gravitational-wave arrows of time.

In Smolin's way of thinking, Newton and Einstein crafted powerful theories by isolating phenomena and directing attention to their geometrical representation ("absolute" space and time for Newton; "spacetime" for Einstein). Smolin calls this strategy "physics in a box." Within the box changes unfold from one instant to the next and the laws of physics render them predictable. Predictions assume the form of mathematically constructed, geometric objects—from lines on Cartesian planes (Newton) to cones in Minkowski spacetime (Einstein). "Clock" time, rather than the arrow of time, works best with physics in a box.

Together, the rules governing interactions within the box and the initial state of conditions within the box determine what changes. That understanding logically entails both past and future states. In these representations time becomes a mathematical construct, the same as all the other variables. Time's arrow fades into the shadows. Mathematical expressions just exist; they need no time to happen. All the changes they stand for are wrapped up in their logic, and logic is the law. Mathematical representation lays bare the stark realities of logic, and logic has no before and after

In nature, causes exist. The sciences seek to know them. Ironically, because mathematics inherits the timelessness of logic, "cause" takes on a strange status. "Time" and "cause" are not synonyms, but causes require time. Order in time is essential to causation.

The geometries and equations that comprise classical and relativistic physics symbolize causation but are not causes. Interestingly, they are no more causal than the expression 2 + 2 = 4. 2 + 2 *is* 4. Two plus two does not cause 4; mathematics has no causes. Very young children make this remarkable leap into abstract logic as they learn arithmetic. Combining two twos entails four by logical necessity, given the definition of counting numbers.

There is no dimension of time in the two-plus-two equation. The equality exists and works equally well left to right and right to left. Analogously, the clock-in-a-box often times things forward and backward equally well: "A law of physics can often be run in reverse . . . a law that can be run backward to turn every final condition into its initial conditions is time-reversible."[4] This time symmetry is a property of the representation—the mathematical analogy used to capture reality. The arrow of time, on the other hand, seems to fly in only one direction. Same realities; two metaphors. They divide thinking in order to solve different problems.

Mathematical objects are not the stuff of nature. Causes exist in nature; logical relationships comprise mathematics. That mathematics represents both everyday and scientific observations of the world so successfully is quite stunning; it's a powerful metaphor. Maybe the universe ultimately consists of information, a bit of reality here, a bit of reality there, metaphorically speaking.

Calendar Time and the Popcorn Clock

Historical sciences commonly employ length as a metaphor for time. For example, Carl Sagan popularized the "Cosmic Calendar" as a representation of cosmological and geological time scales.[5] The Cosmic Calendar expresses time, as do all timelines, as length. Yet time is not literally a line and thus has length only by analogy to space. Time is time, present as a moment, factual as the past, and open to possibility as the future. Time lacks a line's symmetry. Arrows, lines, equations, the common trinity of metaphors for time, fail to do time justice.

"In order to time something, you have to compare it to something that always goes at the same speed," explained the fourth-grader. He continued, "But you cannot tell whether something goes at the same speed unless you time it!"[6] Even a fourth-grader can recognize that timing, like time, feels a bit paradoxical when defined. Measuring geologic time adds confusion by introducing "randomness." The "something" that "goes at the same speed" depends upon random events at the atomic scale. The meaning of time cannot escape association with probability. Because the ways of being disordered are so many and the ways of being ordered relatively few, disorder is the more likely state of things as time goes by. Association with probability gives flight to time's arrow.

Radiometric dating functions as the geologist's clock, a method that historically opened insights into the reality of truly "deep time." Curiously, the individual ticks of the radiometric clock tick randomly. One atom decays, then another. There's no telling which one is next. How does a scientist use a series of random events as a clock?[7]

Over time, radioactive decay exhibits regularity. In a characteristic amount of time, one-half of the amount of a parent isotope decays into new isotopes of a different element. The ratio of these two isotopes is a function of time. A radiometric clock actually ticks in half-lives. Which atomic nucleus may decay stays unpredictable, but the probability of its spontaneously decaying is not. When probabilities are high, half-lives are short, as in carbon dating. When the probabilities of decay are low, half-lives are longer, as with uranium. The logic that exploits a randomized event as a reliable clock is solid, even if tortuous for students to learn. The "popcorn clock" is a useful analogy for introducing radiometric dating.

For twenty years, a middle-school science teacher in Hawaii used popcorn popping as an analogy for understanding radiometric dating. She heated a batch of kernels in a frypan of oil and posed the question, "Which kernel will pop first? Then second?" With guidance, her class accepted the notion that the kernels, all at the same temperature, popped randomly through time. With time, the ratio of popped to unpopped kernels increased—the same as for isotopes of unstable elements.

Anyone who has listened patiently to popcorn popping knows that the popping rate speeds up, partly because temperature keeps increasing. If kept at a constant high temperature, the probability of a kernel popping rises. Some take longer than others to get ready to pop. Radiometric clocks supposedly tick constantly and do not respond to temperature variations. Analogies do have their limits. However, if productive, they start the engine of thinking effectively.

Converting isotope ratios into a clock permits remarkable measurements of the past. Representing time as length helps to visualize and compare durations. Alas, many teachers, as well as the movie *Grand Canyon*,[8] dutifully call up the metaphor of time as length to reinforce a sense of human transience and unimportance. The view from the canyon's south rim does travel backward in geologic time through durations long enough to assemble and disassemble two supercontinents, Rodinia and Pangea.

The canyon is vast and deep, yet even the time to carve the canyon (perhaps as "little" as ten million years) pales in comparison to creating and destroying Pangea. Author John McPhee's metaphorical phrase "deep time" captures the sense of vast duration. Indeed, from the rim of the Grand Canyon, time actually looks deep.

From the metaphorical perspective of earth as Gaia, a living organism, extent in past time recedes in significance. The present moment matters and is pregnant with possible futures, some of which human action might harm.

Tempering the time-as-length metaphor while keeping knowledge of earth's long history intact fosters deep respect for the present moment. Events deep in time remain the foundation of the present, enabling and constraining the future. The present moment contains and reworks all things that have ever happened, with novelty and complexity being the result. Even the most ancient fragment of the earth "is, at each instant, the summary of its past and the threshold of its future."[9]

From this respect, Danny Glover and his friends in *Grand Canyon*—as well as all schoolchildren—may seize on a sense of individual responsibility and be exhilarated by a sense of time. Being present at the threshold of the future is heady stuff. This image overcomes the misconception of temporal insignificance; too often the message reinforced by the time-as-length metaphor. As William Falkner poignantly wrote in *Requiem for a Nun*, "The past is never dead. It's not even past." The past stares from the walls of the Grand Canyon, whispering, "Now what?"

Temporal reasoning permeates the task of interpreting the earth's history and structure. Many of the important concepts in geology encapsulate a time series: its key words imply stories. The reason is simple: every feature of the earth has a history. Consider eruption and erosion, for example. Each of these concepts refers to a process that unfolds in time. Good geologic reasoning depends upon making comparisons in terms of carefully restrained yet ultimately ambiguous categories. All deltas are similar to each other; at the same time, each one is unique. The same observation applies to mountains and faults and even crustal plates. These are the objects of interest in geology, and their present states are contingent on their pasts.

Productive comparisons among deltas and between deltas and seafloor fans yield insights into the processes that formed them, and inferring processes from patterns is the goal of geologic inquiry. In contrast, individual atoms of carbon 12 are utterly indistinguishable and completely interchangeable. A carbon atom does not come wrapped in the history of its past interactions; the erosional phenomena of the Grand Canyon of the Colorado River, do.

Physics names its fundamental entities with definitions designed to enable measurement. Fuzzy boundaries will not do. Mass, energy, force, and temperature are quantifiable properties, and the procedures for recording their values are finitely described. The terms speak in measured tones. Physicists represent changes with mathematical objects. Geometric representations express changes and relationships. Equations rise to lawful status to govern what physicists conceive.

Geologic rhetoric is temporal rhetoric. Its words tell stories. Eruption and erosion happen as a sequence of events. Geoscientists often represent temporal relationships visually, depicting data in forms that promote spatial reasoning. Geologic maps are the perfect—and stunningly attractive—example. They color time.

TADPOLE DESTINY

Darwin departed from his fellow transmutationists, with whom he shared the notion of species change through time and the origins of complex forms from more simple ones, by insisting on natural selection as the principle mechanism of speciation. His mind broke with prior centuries of thought not only because he had collected the evidence to craft a natural science account of life's history in opposition to religious dogma, but also because he appealed to the right metaphors and analogies to form his new idea.

Natural selection operated analogously to selective breeding. It seized upon minute variations—unintended, unanticipated, and uninfluenced by surrounding conditions—to fashion novelty. Pigeon fanciers understood this process, and Darwin recognized the rock pigeon as common ancestor to numerous forms. Selection happened from without, either by someone doing the selecting in the case of domestic plants and animals or naturally in the struggle for existence.

In contrast, his nemesis Jean Baptiste Lamarck, his own grandfather Erasmus Darwin, and Robert Chambers, the author of *Vestiges of the Natural History of Creation*, ascribed to a metaphor of metamorphosis. Non-Darwinian transmutationists imagined that species transformed into more advanced forms in a manner similar to metamorphosis in the life cycle of individual organisms: egg to infant, caterpillar to butterfly, or tadpole to frog. From old species, new and different ones emerged.

For transmutationists, progressive development, as in embryos, responsive to the demands of living, resulted in new forms of life. For Erasmus Darwin this process began as a single filament, the progenitor of all creatures. As species "transmuted," or evolved, into increasingly complex forms, simple forms spontaneously generated out of the muck of decay to fill the depleted lower rungs of the ladder of life.

For Darwin, this was speculative mysticism. On the other hand, modern biology, whose root metaphor is information in the genetic code of life, sees a connection between speciation and mutations in the controls of development. Fins become limbs as evolution tinkers with the information stored in DNA—the true primordial filament.

Darwin's coauthor of the first paper on natural selection, Alfred Russel Wallace, anchored his idea of survival of the fittest as the cause of species to deviate indefinitely from type as an economic theory of population. He as well as Darwin drew inspiration from Thomas Malthus's theory of how diseases and destitution stopped exponential increases in human population growth. The natural world, they concluded, functioned analogously. Individuals varied; selection winnowed the variants. Survivors became the new population.

In keeping with the Malthusian analogy, populations over time—not individuals within their lifetimes—adapted to the conditions of life in a struggle for existence. The ever-branching, Malthusian-rooted tree of life supplanted the imagery of metamorphosis in the origin of species. Disease, predation, and starvation—not an inner drive to perfection—did the hard lifting.

METAPHORICAL ALCHEMY

Jacob Bronowski, the historian of science and author of the 1973 BBC program and book *The Ascent of Man*, famously wrote, "Every theory is based on some analogy, and sooner or later the theory fails because the analogy turns out to be false. A theory in its day helps to solve the problems of the day."[10]

Bronowski found good things to say about the imagery that guided medieval alchemists. They believed that the principles governing the body operated similarly in nature. Mineral, vegetable, and animal kingdoms followed the same rules of transformation. According to the alchemists, minerals as well as the traditional herbs promised palliative effects on human diseases; mercury as a treatment for syphilis was one result.

Physicists ultimately brought the heavens into this fold: cannon fire and planetary orbits conformed to the same rules. Smaller bodies orbited larger ones, analogous to how the moons of Jupiter circled their giant host. The moon, a celestial object, was as corruptible as any temporal, earthbound one. Heavenly objects—eternal, perfect, and incorruptible—were not so heavenly after all. They changed through time just like the earth. Continuous creation and inevitable deterioration united both realms. Increasingly, the sciences realized that the same laws and principles governed body, nature, and heavens—perhaps even mind at some level. Bronowski might have written that sooner or later a better analogy comes along to break a barrier and advance the science.

Metaphorical imagery undergirds the diversity of the sciences in subtle as well as obvious ways. The process of finding metaphors and evaluating their consequences starts the diversity search engine. Physical sciences' "particles" (both dimensionless and space-filling), Darwin's "tree" of life, astronomy's "black holes," chemistry's atomic "orbitals," cosmology's "big bang," meteorology's "atmospheric windows," geography's rain "shadows," geology's "plate" theory of tectonics, cognitive science's "hardwired" brain, and science education's "pipeline" are just a few. Pronghorn males keep a "harem" of mates; bees serve a "queen." Is the earth an organism? Is the body a heat engine? Is the brain an information processing system? The list of metaphors serving explanatory purposes across the disciplines ranges widely. They open doors.

The abstract lists of generic NGSS practices, crosscutting concepts, and nature of science (NOS) understandings fail as a context for learning core disciplinary ideas. Tangible, disciplinary problems work much better when well-chosen metaphors and analogies make these contexts accessible. It is difficult to underestimate the significance of imagery to the diversity of scientific thinking. Metaphor and analogy join the distinction between styles of science and the example of geologic reasoning to complete a short list of starting points for honoring disciplinary diversity.

The generic viewpoint explicitly presented by the NGSS eclipses insights into how the sciences adapt to a myriad of contexts from their metaphorical starting points. Metaphors, analogies, agendas, methods, and conceptions differ. They must, in order to get the work done. The fundamental flaw in the NGSS, more than a half century in the making, is unsurprising given widespread acceptance of the underlying imagery of the quest for unity.

However, the dream of unity has run its course and the imagery guiding reform needs to change. Efforts to place the tools of the trades within the grasp of novices would benefit reform-minded thinkers, with "tools" meaning both the inventions of mind and the techniques of investigation engineered to match them. The trades are many and their tools diverse, and that's the imagery needed to solve science education's "problems of the day."

Beneath a famous drawing of a pipe, the French surrealist artist Rene Magritte wrote, "*Ceci n'est pas une pipe.*" The translation is: "This is not a pipe." The caption and image combine to mean that the drawing only represents reality; reality is not the drawing. Disciplines depart from each other in how to represent salient features of reality. They draw different pipes. Students, in turn, study these representations.[11] They use analogies and metaphors to draw reality into the mind.

NOTES

1. Douglas R. Hofstader, "Analogy as the Core of Cognition," in Dedre Gentner, Keith J. Holyoak, and Boicho N. Kokinov, eds., *The Analogical Mind: Perspectives from Cognitive Science* (Cambridge, MA: The MIT Press/Bradford Book, 2001): 499–535.
2. Lee Smolin, *Time Reborn: From the Crisis in Physics to the Future of the Universe* (Boston: Houghton, Mifflin, Harcourt, 2013).
3. Stephen Jay Gould, *Time's Arrow, Time's Cycle* (Cambridge, MA: Harvard University Press, 1987).
4. Smolin, *Time Reborn,* 51.
5. Carl Sagan, *The Dragons of Eden: Speculations on the Evolution of Human Intelligence* (New York: Random House, 1977): 15.
6. Charles R. Ault, Jr., "An Elementary Conception of Time," *Thinking: The Journal of Philosophy for Children* 6, no. 4 (1986): 7.

7. Radiometric dating helps order events in geologic time. It is not the same as the atomic clock used to define a standard unit of time. In an atomic clock, an excited cesium atom emits microwave radiation at a precise frequency and is therefore a dependable constant. Cesium electrons emit energy as they transition from excited to resting states. Radiometric dating depends upon decay events in the nucleus of an atom.

8. Lawrence Kasdan and Meg Kasdan, *Grand Canyon* (Hollywood, CA: 20th Century Fox, 1991): film.

9. David Leveson, *A Sense of the Earth* (Garden City, NJ: Natural History Press, 1971): 129.

10. Jacob Bronowski, *The Ascent of Man* (Boston: 1973): 141.

11. Rosalind Driver, Hilary Asoko, John Leach, Eduardo Mortimer, and Philip Scott, "Constructing Scientific Knowledge in The Classroom," *Educational Researcher* 23 (1994): 5–12.

3

PROSPECTIVE SOLUTIONS

Seek to engage the natural world in all of its variety and innovation.

—Deborah Seltzer-Kelly[1]

School science aims to represent what scientists do while preparing citizens for lives of democratic living. Diverse approaches to scientific inquiry in school curricula best reflect what scientists do, not a quest for unity. Conceptions of phenomena winnow relevant information from irrelevant information and assign meaning to observations. Conceptions offer coherence whether learning a traditional subject or applying disciplined thinking to solving community problems.

Abandoning the quest for scientific universals in order to frame science for all opens a cornucopia of choices. Asking the disciplinary question, "Which conceptions best organize thinking in well-defined contexts?" and the citizen-science question, "Which conceptions offer the most value to preparing for democratic living?" constrain the cornucopia while accepting the need for diversity throughout the K–12 curriculum.

NOTE

1. Deborah Seltzer-Kelly, "Deweyan Darwinism for the Twenty-first Century: Toward an Educational Method for Critical Democratic Engagement in the Era of the Institute of Education Sciences, *Educational Theory* 58, no. 3 (2008): 304.

11

The Twin Pillars of Thinking and Doing

According to the nineteenth century's "biogenetic law," a powerful embodiment of Darwinism, an organism's development passed through stages sequenced in the same fashion as its evolutionary descent. In the jargon of its era: "ontogeny recapitulated phylogeny." For many, being a Darwinist meant being a recapitulationist.[1]

Thus, recapitulationists argued that the development of an organism, from embryo to adult, repeated its biological past. Growth mirrored descent. For example, even human embryos sport gills for a while, evidence of humanity's piscine ancestry. Recapitulationists further explained that innovative adaptions added during the course of adulthood migrated into earlier stages of development.

Surprisingly, the biogenetic law is a fundamental driver of the quest for unity in science education. The law descended with modification by means of its selection in the thoughts of Jean Piaget and John Dewey. Piaget's research reinforced universal notions of reasoning; Dewey's thinking encouraged generic approaches to inquiry.

Their ideas also directed attention to crafting child-centered experiences, not just imparting information in the classroom. They emphasized the importance of thinking about the thinking. Despite their role as the twin pillars of unity in science education, there is a great deal of Piaget and Dewey that offers promising solutions to the problems brought about by the quest. Examples of diversity in classrooms boasting a child-centered approach and engaging students in real-world problem-solving are easy to find. Among them are a school predicated on teaching sustainable living, a community dedicated to conserving leatherback turtles, and children immersed in "minibeast" study.

THE IMAGERY OF RECAPITULATION THEORY

Recapitulation imagery found striking resonance in the works of Jean Piaget and John Dewey. Piaget characterized stages in the growth of intellect as biological adaptations aligned with the progress of human civilization. Dewey believed that a child's inquiry should recapitulate the experiences of the original investigator. Their notions of thinking and doing became twin pillars of science teaching; Piaget suggesting the ends and Dewey the means. In simplest terms, their teachings proclaim, "Inquiry experiences develop intelligence!"

Belief in the equal applicability of evolutionary principles to nature and society led nineteenth-century educators to extend the biogenetic law to child development. A leading proponent of this movement, G. Stanley Hall, for example, concluded that the child must recapitulate the history of society in order to develop properly.[2] Recapitulation theory (the biogenetic law extended to human learning), leading voices supposed, would put educational theorizing on a sound scientific basis.

They espoused a curriculum that mimicked the sequence of historical periods as understood by Western cultures. The child's intellectual development presumably paralleled the same epochs in the same sequence. Therefore, the task of curriculum design became matching school experiences to the child's developmental stage.

History defined these stages: hunter-gatherer pre-civilization, agricultural society, Biblical patriarchs, chivalric feudalism, industrial revolution, and modern science and mathematics. A program of instructional experiences matched to these epochs promised to develop the child's physical, emotional, moral, and intellectual potential.

The biogenetic law unified disparate disciplines. Thinking in terms of recapitulation linked social and psychological theorizing to basic scientific understandings of evolution. Using history to sequence schooling promised to optimize child development. However, both Piaget and Dewey rejected the idea of education mirroring historical epochs in lockstep fashion.

Piaget accepted recapitulation, but argued that the child's development recapitulated not historical epochs but rather the history of advances in human intellect. The child repeated these advances in progressing from sensorimotor reasoning to formal logical operations. Piaget viewed logical reasoning as an extension of biological adaptation—an advanced stage of development, the adult mind derived from its immature stages.

Dewey also rejected the idea of basing curriculum rigidly upon historical epochs. Instead, he examined inquiry through the lens of recapitulation. The learner, according to Dewey, needed to progress through stages of experience and insight analogous to those felt by the original inquirer. The learner's own inquiry, stemming from personal interest and given direction by different subjects, provided the experiences needed for bridging the gap between expert and novice. Done well, inquiry aligned vital components of original inquiry with a child's development.

Dewey argued that a student ought to experience (in effect, recapitulate) the logic of action pursued in the course of original inquiry. For Dewey, an episode of inquiry in the curriculum recapitulated the important stages of the original.

Piaget's and Dewey's concepts of adaptation extended Darwinism. For Piaget "mental operations" (e.g., seriation, classification, propositional logic) adapted human interaction with the world. For Dewey investigative methods evolved as well. Piaget's stages of intellectual development (from sensorimotor learning to formal operations) and Dewey's focus on inquiry extended the Darwinian idea of recapitulation beyond the realm of embryos. Both rejected the confusing array of historical epochs as the organizing principle of curriculum in making this extension. The logic ("concrete and formal operations") in Piaget's theorizing and the pragmatic approach to "scientific inquiry" in Dewey's thought undergird science education even today.

In their early twentieth-century writings, Piaget and Dewey recognized the crucial role of experience to learning and development. For Piaget, the child's engagement with surroundings richly provisioned with challenges propelled advancement from perceptually bound logic to abstract thinking. For Dewey, inquiry, initiated by personal interest, extended the child's experiences into the wider world by calling upon disciplinary resources. "Richly provisioned surroundings" and "disciplinary resources" deserve center stage in twenty-first-century teaching respectful of conceptual and methodological diversity.

RECAPITULATING THE LOGIC OF INQUIRY

John Dewey's philosophy transcended the pitfalls of recapitulation theory mired in the march of children through historical epochs. He accepted the need to match experience to a child's stage of development. However, for Dewey the guidance offered by cultural epochs led to a feeble dead end. He understood epochs in the history of Western culture as resources for curriculum development, not an *a priori* sequence of study. The present epoch offered compelling problems of democratic living, problems posed by advances in technology, and problems for modern science to address. It had its own problems to solve.

Dewey lamented the preponderance of mind-numbing, stultifying, authoritarian schooling that characterized children's experiences in the early twentieth century. His critique focused on the troubling separation of children's experiences and interests from school subjects. The subjects taught at school lacked problems to solve and fascinating questions to answer. They needed, Dewey concluded, a context of inquiry, a context that might re-create the pathway to insight traversed by original thinkers.

Dewey acknowledged the value to learning of aesthetic appeal, social interaction, and personal interest. He believed that learning required a problem to be made one's own. He lamented that even though resources for solving problems abounded in the subjects taught at school, accessing them was often a dreary and rote process.

The logic of a discipline was rarely accessible to the psychological development of the learner.

The mismatch, thought Dewey, was the consequence of a deficit of relevant experience—including aesthetic and social experience—at the personal level. Schooling's haste to teach formalisms without imbuing them with relevant experiences inflamed the mismatch. Dewey criticized traditional schooling as a place where only the most docile and obedient could affect an air of mastery.

In his 1902 essay, "The Child and the Curriculum," Dewey warned of the consequences of failing to integrate children's experience with school subjects. He argued that the words and concepts of science were "dead and barren" unless connected to the child's experience. "The lack of any organic connection with what the child has already seen and felt" and the "conspicuous absence" of "need and aim" for the subject matter become a miseducative evil: "Those things which are most significant . . . and most valuable in the logic of actual inquiry and classification drop out."[3]

Seeing knowledge as purposeful was the key. If the purposes of the expert remained opaque, then the learner had no real chance to see the value of the knowledge. Better to collect, culture, and contrast snails from stream and pond so as to gain insight into species diversity than to memorize their taxonomic classification to no end.

Traditional schooling sanitized learning by offering only sterile experiences—categorization schemes bereft of purpose. At school, children failed to experience the aesthetic appeal, the social interaction, and the tangible experiences known viscerally to subject experts. Dewey argued that through inquiry initiated by the child's personal interests and amplified by disciplinary resources (concepts and skills) these dimensions became accessible.

History for Dewey was no dictator of curriculum design. He parted company with G. Stanley Hall and those who would sequence child study to mirror historical epochs. What, then, would Dewey have children recapitulate if not "the history of the race"? The answer is: *an original act of inquiry*. Dewey invested the idea of recapitulation in inquiry. The child, according to Dewey, needed to progress through stages of experience and insight analogous to those felt by the original inquirer. The child's own inquiry, stemming from personal interest given direction by problems derived from different subjects, provided the experiences needed for bridging the gap between expert and novice. Done well, inquiry aligned vital components of original inquiry with a child's development.

For followers of Dewey, what to recapitulate has remained ambiguous. Should an inquiry lesson embody the basic processes supposedly common to all the sciences or the particular methods adapted to the demands characteristic of a discipline's problems? Representing the diversity of scientific methods favors the latter point of view; the quest for unity, the former. In either event, the link to Darwinism by means of extrapolating the biogenetic law remains. Pragmatic solutions—the Deweyan ideal—equate with survival. They are successful adaptations.

Teachers incorporate a bit of Darwinism in every inquiry lesson they fashion. Educators need to guard against the reduction of Dewey's conception of inquiry to the scientific method of school discourse. As educational researcher Deborah Seltzer-Kelly aptly states, "Dewey's perspective . . . was firmly grounded in his perception of scientific method as it actually exists in the sciences, which in turn was grounded in Darwin's evolutionary theory."[4] Dewey valued "scientific method as it actually exists in the sciences." There is a vital dimension to this perspective that respects the diversity among problems and the variety of methods adapted to solving them.

In Dewey the biogenetic law has found safe haven: the child must recapitulate the experiences of inquiry through inquiry. For Dewey, child-centered study reenacts crucial experiences and insights of original inquiry. Whenever and wherever a K–12 teacher succeeds with an "inquiry science lesson," the ghost of Dewey smiles. The challenge to educating is to translate original inquiry into school curricula without losing the psychological aspect of purpose. Dewey's conception of inquiry, properly interpreted, reinforces the importance of diversity among the scientific enterprises.

The following two vignettes describe how Deweyan principles may unfold in the twenty-first century. Both are examples from rural settings—an alternative high school in the United States and a community project in Costa Rica. Neither case represents mainstream education, and that is their strength. If thinking about schools is to change, some examples from the margins must be considered.

School of Sustainability

Natural selection has fashioned apple blossoms just right for pollination by bees. Orchards need bees—so do clover fields, broccoli gardens, and almond trees. The past decade has seen bees in peril and an increase in the need for attending to bee husbandry. The bee population's role in food production is a national concern. Climate change, to the extent that it may affect disease vectors, may be among the culprits responsible for bee decline.

Secondary students at the Atfalati School maintain four honeybee hives on their campus. They host elementary students who come to learn about pollination and ecology. In addition, the high-school students present workshops to members of the local community on apiary management and beekeeping as a backyard hobby. Atfalati students also harvest honey, of course. Theirs is a very sweet curriculum indeed.

The Atfalati School maintains a community garden, teaches watershed conservation, and asks students to think deeply about food security. Sustainable agriculture is a practice central to the school's curriculum. So, too, are boatbuilding, forestry, and architecture.

Boatbuilding began as a student interest project. First came scale drawings. After several months, a small team had successfully constructed an Aleutian double kayak using Douglas fir gunwales, oak ribs, and nylon skin. They appeared to have succeeded, but a maiden voyage up Coyote Creek had yet to happen. A female student, who had remained after school on occasion to add some finishing stitches to the skin, declined an invitation to sit in one of the cockpits, saying, "I don't like being in a boat controlled by one of the boys."[5]

"Archaeologists have found evidence that shows that these kayaks go back over 2,000 years, but trying to find evidence is hard because they were skin-covered, sinew lashed, and wood framed so they don't last very long," noted one of the student boatbuilders. "I didn't know that there was so much history in these boats, but they are amazing crafts." His kayak partner added, "The traditional boats were built with whale bones and walrus skin."

Atfalati students collaborate with local landowners with an eye to managing forested lands for sustainable use. They put trigonometry to work in figuring out easement boundaries and work with local conservation agencies on management plans. The curriculum requires them to learn about the complexities surrounding conservation easements, land covenants, and trusts.

Teachers at the school bring valued expertise. For example, two teachers, a former bioengineer and an industrial chemist, facilitate project work with the school's community partners.

Eugene architect Nir Pearlson has worked with Atfalati teachers and students to help them understand the essential systems of green buildings: natural ventilation, rainwater harvesting, micro wind turbines, and living roofs. The school ardently strives to be a school of sustainability.

The Atfalati School does not serve mainstream students. It's an alternative high school, populated with students for whom the conventional classroom was a dead end. Dewey believed that every epoch and place had problems to solve. The Atfalati School addresses two of them head-on: the failure of secondary education to serve all adolescent learners and the need for society to educate the next generation in practices of sustainable living. Dewey's conception of democratic living meant entrusting problem-solving to citizens in partnership with experts. There is no "deferred citizenship" at Atfalati; kids are citizens and community partnership is the norm.

Conserving Turtle Eggs

Midnight has long since passed. The rains are heavy and the surf high; the tide is in. It is extremely dark. In a matter-of-fact way, a group of Costa Rican children assemble around the dining center table as the turtle-watch shift comes in. No new leatherback—*baula* in Spanish—nestings to report; no encounters with turtle egg poachers, either. There remains other important work to do.[6]

Miguel brings in a plastic tub with several leatherback hatchlings, their rubbery flippers rhythmically propelling them around the tub and over each other. Midnight hatchings on a dark night are unusual; these came from the nursery watched over carefully by the local residents of Parismina, north of Limón on Costa Rica's Atlantic coast. It is a roadless area where regenerating forests are traversed by canals and the common occupation of hunting turtles and harvesting their eggs (long considered a delicacy and aphrodisiac, and consumed in local bars) has given way to turtle conservation.

Residents refer to this section of the coast as Estación Las Tortugas (the Turtle Station). Those who visit think of this laboratory of conservation and education as "Stanley's beach," thus honoring the local Costa Rican landowner and Boy Scout leader Stanley Rodriguez Mendez, whose dedication to children and the protection of leatherbacks have inspired the locals and anyone else who finds his or her way to Estación Las Tortugas.

As if by habit, the children, led by thirteen-year-old Teresa, get out calipers, tape measures, and scales. They carefully measure, weigh, and record observations of behavior (for example, how vigorous the crawl) for each hatchling. Observing are a group of teachers from Oregon and their hosts, faculty from the School of Education at the University of Costa Rica. Eventually everyone has a role to play, though several of the teachers must depart with the next shift to patrol Stanley's beach where, on that night, they will encounter a solitary green turtle probing the leatherbacks' turf.

Lightning strikes toward the horizon lift the darkness as the patrol gingerly moves south. The patrol leader's light shines dimly red. Should a turtle be encountered there is delicate work to be done and little time to spare. No other lights are permitted—they might confuse and disorient a female leatherback intent on digging a nest and laying eggs. Shore lights, especially those of resorts, are just one of the human insults leatherbacks never had to worry about for tens of millions of years.

Part of the work of the patrol is to discourage poachers. Predators are not new, of course; but the appetites of men in search of virility and feral dogs scavenging for an easy meal are. The simple presence of volunteers on patrol, night after night through the nesting season, works very well. Feral dogs must be scared away, for they sniff out nests and can rapidly destroy an entire clutch of eggs. Human poachers, who hope to make a living selling the eggs in bars as aphrodisiacs, must be discouraged. The fecundity of sea turtles is striking, and so a belief in the relationship between virility and their eggs has evolved.

The count, behavior, and vital statistics for tonight's hatchlings are crucial to evaluating the success of turtle conservation at Parismina. The local children aged eleven to thirteen collect this data, and participation in this work is essential to their education. Children, after all, are the cornerstones of true work in conservation.

Turtle stats in hand, Miguel, the children, and several very wide-eyed teachers from Oregon proceed to the beach. The hatchlings are released a proper distance from the surf as if they had just crawled out of the nest. The gentle light of the horizon on the splash of the surf attracts the attention of these tiny creatures. A few are destined, once mature, to become behemoths, returning from their sojourn at sea to this or a nearby beach.

The children of Estación Las Tortugas, some of them barely literate, are engaged in research science on a scale that would be the envy of every inquiry-oriented science teacher. Hope for the *baula* resides in children's care and the commitment of coastal communities to conservation practices that make local economic sense. An ember of democratic living glows on a remote Costa Rican beach where scientific practices thrive. Children engage in conservation, creating for themselves the same insights and experiences felt by research biologists. They have achieved a Deweyan ideal where inquiry and survival are linked.

RECAPITULATING THE DEVELOPMENT OF REASONING

The analogy between individual development and species descent encapsulated in the biogenetic law had a striking impact on the young Jean Piaget who, as a zoologist, studied snails. Later he became famous as a psychologist who examined the development of intelligence. Metaphorically, the child-as-snail (or equivalently, the mind-as-organism) shaped his theorizing.

Piaget was neither a time-traveler nor an accomplished anthropologist, yet he believed in the importance of understanding the psychology of prehistoric humans, not just modern ones. People, he assumed, could become smarter if they understood the origins and progress of thinking. He turned his biogenetic attention to stages in the progress of logic and reasoning. Lacking prehistoric humans to study directly as exemplars of the origins of intelligence, he searched for the perfect surrogate and eventually found one: any living child would do.

> Unfortunately, we are not very well informed in the psychology of primitive man, but there are children all around us, and it is in studying children that we have the best chance of studying the development of logical knowledge, mathematical knowledge, physical knowledge.[7]

Piaget first studied mollusks—pulmonate pond snails to be precise (snails with lungs, for example, of the genus *Lymnaea*). He was curious about their adaptation to different habitats at different elevations in Switzerland. He experimented with snails by moving them between lake and stream water and observed snails in ponds at different elevations in the Alps. Piaget found that snail shells morphed in response to changing conditions.[8] Tracing such changes across natural, migrating populations unveiled new species. He concluded that the adaptations made by migrating species led to the origin of new species. For Piaget, adaptation, in effect, extended embryological processes throughout life. Adaptations reorganized existing structures in response to the demands of new environments.

Evolutionary biology had clearly established in Piaget's day that embryos, through reorganization of existing structures, added stages to their development. Most importantly, passing through a stage successfully was prerequisite to subsequent development.

Snails illustrated this principle well. By 1918, before Piaget tried to learn about prehistoric humans by studying children, he had girded his loins for intellectual battle with a deep grasp of gastropod adaptations and coiled himself within the protective shell of recapitulation theory. Recapitulation was his guiding metaphor as he turned his attention from snail adaptations to human ones.

"Piaget assumed that direct adaptation takes place under the pressure of environmental changes."[9] He concluded that the structure of snail bodies adapted appropriately to stressful conditions—to external properties. For example, snail size and snail color changed in response to water temperature and clarity: bigger snails in colder water; lighter snails in clearer water, the better to survive. Today these changes in shell morphology are presumed due to alteration in the regulation of gene expression based on environmental conditions if not because of differential survival of offspring according to size and color.

For Piaget, the growth of intelligence was the continuation of this basic biological phenomenon: the internalization of external properties as useful adaptations. Minds adapted by reorganizing existing structures in response to problems in the same manner as snail shells adapted to their environments.

Consider that the amount of snail remains the same whether the snail is hiding in its shell or stretched out, racing across a tabletop. "Nonsense," the young child might say. The snail may understand this fact quite well, but a young child, still grappling with sensorimotor learning cued by perception, will likely focus on the snail getting bigger and smaller, unaware that if it is served as escargot, the meal would be the same size regardless of coiling and stretching.

Not until properly challenged—repeatedly—can a young child's mind recognize that an amount of clay—or snail, in and out of its shell—remains unchanged by flattening, stretching, or rounding. Stretching increases length, but not amount. "Amount" is a more abstract idea than length. Through interaction with the environment—by making patties of modeling clay—the child comes to internalize physical properties such as amount.

Being able to mentally conserve a quantity of material avoids being fooled by appearances. A mental structure corresponding to constancy in nature gives a thinker the ability to manipulate objects in thought and a craftsperson added skill in fashioning tools. These are adaptations useful to human survival. Once the child has mastered the conservation of matter, acquiring even greater intelligence may proceed. Adapt and survive; evolve and progress. For humans, it's all in the mind.

For Piaget, the development of individual intelligence mirrored the ascent of rational thinking in human history, evolving ever upward toward scientific reasoning. The Piagetian child advances through the accumulated wisdom of the ages with no need to dally in feudal or other historical times. However, without sufficient exercise, mental development arrests at an early stage. Adaptation halts, truncating intelligence.

Analogy propelled Piaget's thought. His theory of mental structure rested upon the adaptability of Swiss snails. Darwinism lurked in the background, gently bending child psychology in its direction. Grasp of logical relations came to rule the formally adapted, Piagetian mind.

Individual intelligence advances stage-wise, teaches Piaget. It adds complexities to primitive states in order to reach higher mental functions. The developing mind traverses the territory mastered historically by the progress of human societies, from tracking prey and finding roots to designing cathedrals and navigating rockets. The young child is Huck Finn; the young adult, Sally Ride.

For the past few decades, nearly every early childhood educator and primary-grade teacher in the nation has learned about Piagetian stages of child development. Secondary teachers, especially of physics, have learned to think of themselves as responsible for helping students transition from concrete to formal operational—and hypothetico-deductive—reasoning. Virtually all American children in some way or another are affected by how teachers enact the Piagetian version of the biogenetic law.

"Childhood is a biologically useful phase whose significance is that of a progressive adaptation to a physical and social environment," he wrote.[10] His work epitomized the origin of promising thought from strange antecedents—culture-epoch theory, the biogenetic law, and Swiss snails.

Piaget credited Socrates's emphasis on a pupil's own activity rather than school-induced docility as a valid precursor to his own philosophy. Piagetians have underscored the importance of schooling that "appeals to real activity, to spontaneous work based upon personal need and interest"[11]

Piaget substantially advanced the understanding of how the thought and inferential capacity of children differed from that of adults. Several Piagetian insights continue to hold even if his stage theory—and its formal operations pinnacle that makes physics the prototype of scientific reasoning—becomes suspect.

Minibeasts

Science 5/13, a British curriculum of the 1970s, intended to show teachers how to help children from ages five to thirteen "learn science through first-hand experience using a variety of methods." The authoring team set forth objectives for the program explicitly in Piagetian terms and outlined stages from "intuition to concrete operations" in the "transition to abstract thinking." For each stage, the curriculum outlines nine categories of broad aims. One broad aim is to develop "aesthetic awareness"—a dimension to learning science often neglected.

"Ugh, here's a slug—shall I kill it?" So begins the introduction to *Science 5/13*'s *Minibeasts*.[12] A minibeast is any small, invertebrate creature inhabiting the natural world familiar to children and close to home and school: soil, garden, woodland, field, stream, and pond. Some do have fierce names, such as water bears and dragonflies. Children observe minibeasts in the wild or in captivity in their investigations.

On page one the authors share their primary objective for *Minibeasts*: children are expected "to acquire a sensitivity to the need to give living things proper care." A context of value comes first and, of course, slug-icide is rejected. Attention to values and aesthetics frames *Minibeasts*; Piagetian theory leads from behind, and "first-hand experience" sits at the top of a diagram of the curriculum. The guidelines for teachers urge them to start lessons with examples that capture interest—hawk-moth larvae or perhaps a large, hairy spider, for instance. The guidelines do date the materials, suggesting that teachers worried about handling creepy creatures can count on "at least one boy in the class who will take charge." That was 1973-speak.

Minibeasts also suggests utilizing community experts. For example, a beekeeper might be invited into the school to share samples of honeycomb. Prompted by first-hand experience, investigations may follow. How strong is honeycomb? Is beeswax a good polish? What does pollen look like under magnification? Where does honey in local stores come from? How is honey used in cooking?

As another starting point, the authors recommend introducing children to snails and worms and then constructing wormeries and snaileries to husband them through a couple of generations. Worm castings and pearly snail eggs appear in short time at the top of the soil. Observations of external anatomy, aided by a hand lens, inevitably reveal surprises, as does watching these minibeasts from below as they crawl over a clear, Plexiglas surface. "Look! They slimed it!"

Children learn to abstract ideas about behavior and life cycles from comparing minibeasts. Mealworms are beetles that pupate, as do moths. Adults feed differently than juveniles. Slugs are very similar to snails, but lack shells. Earthworms are hermaphrodites; so, too, are many pulmonate snails.

How minibeasts move and what they do upon encountering objects in their natural habitats may capture and sustain interest for extended periods of time. How do snails differ from slugs as they move through their habitats? Slugs cannot withdraw into a shell. Do they produce more mucus than snails?

The study of a group of minibeasts—Swiss snails—inaugurated Piaget's scientific career. He was a master of minibeast science. Many of his questions were similar to those found in the lessons of *Minibeasts* that focus attention on anatomy, behavior, and the struggle to survive. Piaget's interpretation of snail adaptations echoed throughout his theorizing on child development. In turn, the *Science 5/13* authors framed minibeast study in terms of Piagetian stages of intellectual growth.

A most notable, non-Piagetian precursor to *Minibeasts* exists. First written in 1911, Anna Botsford Comstock's classic, *The Handbook of Nature Study*, presents 232 unsurpassed lessons on plants, animals, earth, and sky. Journaling and field study are essential methods in Comstock's curriculum and she correlates nature study to language, history, geography, arithmetic, gardening, and agriculture.[13]

Comstock's questions guide children's inquiry in the same fashion as the *Minibeasts* authors encouraged decades later. Neither truly depends on Piagetian theory to generate instruction. In "Lesson 106: The Garden Snail," Comstock instructs teachers in how to build a snailery. Prompted by rhyming couplets such as "With your eyes on stalks, and a foot that walks," children observe the snail climbing and clinging to the glass wall of the snailery. The imagery is more memorable than the steps of the inquiry jingle.

Comstock draws children's attention to the track of mucus. Her robust set of questions includes, "Where are the snail's eyes? Can it pull in one eye and leave the other one out? . . . How many spiral turns are there in the full-grown shell? . . . Can you find the opening through which the snail draws its breath? . . . Place a snail on its back and see how it rights itself." Comstock, too, encourages children to look for nearly transparent, glistening snail eggs fastened together by mucous.

Background on gastropods in general as well as slugs in particular precedes "Lesson 106." Comstock's "Lesson 107: The Earthworm" follows, calling for children to make comparisons among invertebrate forms. Comstock explains, "In nature-study the work begins with any plant or creature which chances to interest the pupil." Her *Handbook* includes a lengthy lesson on plant lice (aphids) and several on spiders and their types of webs. Curiously, the *Handbook* omits the ubiquitous crustacean, the pillbug.

Isopod-query

Pillbugs and sowbugs (two groups with the order *Isopoda*—often referred to as roly-polies, potato bugs, or wood lice) scurry about in decaying material such as compost, woodpiles, and leaf litter. They congregate but do not live in hives. Do they respond to each other upon contact? What odors attract them? Often children will observe a gel-like patch on a pillbug's abdomen. It's a brood pouch and young are born segmented and multi-legged, ready to walk and roll. As one child exclaimed, "I can see through the baby's body!"

With guidance, children may invent investigative methods for testing how different isopods respond to encounters with each other, different creatures, and environmental stimuli.[14] Pillbug responses to food, temperature, light, odor, and moisture preferences are amenable to study by children. Isopod-query teaches them to match their methods of inquiry to the demands of the question. Their investigations give context to what Dewey referred to as the "logic of action" and Piaget to "concrete operational" thought. Procedures for investigating the behavior of isopods readily transfer to the study of other minibeasts.

For example, isopods may prefer to crawl about within a particular range of temperature. To explore this idea children in a sixth-grade science class scattered two dozen sowbugs (isopod flatteners, not rollers) across a long tray filled with wet sand, one end resting upon a bag of crushed ice and the other upon a thermos bag of hot water. They covered the tray with Plexiglas marked with lines that divided the tray into six sections from cold to warm. The Plexiglas cover also kept the humidity high throughout the test habitat. After a few minutes had passed, they took turns measuring the temperature and counting the number of sowbugs in each section.

The teacher first asked, "Do the sowbugs like to be warm or cool?" Answering depended upon finding more isopods at one end of the tray than at the other. Of course, questions progressed, for instance: "What would count as evidence that the isopods preferred a particular temperature?"

Answering this question depended upon finding more isopods in a section than would be expected by chance. Interpreting the data ranged, from a Piagetian perspective, from thinking in terms of association and perhaps a trend—a concrete reasoning task—to wrestling with distinguishing non-random distribution from a random one, a formal reasoning task that might tax students in an introductory-level college biology class. Formal reasoning does have precursors.

There are many investigations to conduct utilizing the temperature gradient tray. Its design controls moisture (at least crudely) and surface texture. The setup also controls light. The experiment is best conducted in a dimly lit room with just enough red light to make observations (pillbugs are blind to red light). One comparison to consider: Do pillbugs (the ones that roll up) prefer the same temperature as sowbugs (the ones that flatten their bodies when threatened)? If there are many newborns in the pillbug colony, children may wish to compare juvenile to adult preferences.

The apparatus for isolating variables and testing their effect upon isopod behavior suggests how to study the behavior of other minibeasts—snails, earthworms, and mealworms. Complexities do multiply, especially when two factors interact. Does seeking a cool place override trying to stay moist? The key point for children to grasp, once again, is how to match the methods of investigation to the demands of the problem. They may choose to test food preferences rather than environmental factors. In any event, good teaching helps them to refine their designs and keeps the logic for interpreting results within the range of children's intellectual grasp.

Temperature may be held constant throughout the sections of the tray, and moisture varied from one end to the other. Children can investigate isopod preferences in a much simpler, dichotomous design: a saucer with one half moist, the other dry, or one half warm, the other cool. The dichotomous design—one-half shaded, one-half lit—works well when testing isopod response to light. Where do they congregate? Generally, the answer is: moist, cool, dark places.

Isopods scamper and burrow outdoors. "Where do isopods live?" is the natural question. "Do pillbugs and sowbugs live in the same places?" Sometimes children notice a lonely pillbug strolling across a dry sidewalk. More often, they find a whole bunch beneath a flowerpot where it's cool, moist, shady, and redolent of compost.

Planimals

Planaria (freshwater flatworms) also make good subjects for study. They inhabit stones in streambeds and appear as dark blobs. Second-grader Robbi learned to set out cans with scraps of liver as planaria bait. He was quite successful.

Crawling reveals their arrow-shaped head with eyespots that appear to cross. Anna Botsford Comstock would no doubt ask children, "How do planaria orient themselves to running water? What do they eat? Which animals are more common in moving water, clinging to stones, and which in calm water, resting on mud? Do planaria live in both worlds?"

Children rather easily classify wormish planaria as animals. However, planaria's regenerative powers fascinate anyone who has studied them. This flatworm can regrow lost sections of its body, including its head (or what passes for a head on a worm). Even ancient naturalists recognized the ability of a few animals to regenerate amputated body parts—starfish arms, for example. Planaria carry the animal example to an extreme, impinging on the regenerative capacity of plants capable of growing entirely new individuals from small cuttings.

Another freshwater minibeast, the tentacled "hydra," a polyp named for the mythical beast that regrew two heads for each one severed, stunned eighteenth-century naturalists with its regenerative abilities. In 1740 two French children, Antoine, age six, and Jean, age three, accompanied by their tutor, Abraham Trembly, collected samples of water from the ponds and ditches of their country estate. Trembly, who adhered to the maxim "ideas should be preceded in the mind by curiosity,"[15] practiced "object lessons"—lessons that focused attention on an object or phenomenon of high interest.

Antoine and Jean observed the tiny creatures contract into little dots and then extend their arms—or were they branches? Hydra walked like animals and regenerated like pruned plants. Cut in half, two individuals grew back—as if the creature were immortal. Hydra and planaria once strained naturalists' definitions of living and nonliving, plant and animal. To Antoine, Jean, and their mentor, hydra and planaria appeared to be plant-animal hybrids, or "planimals." Trembly and his charges shipped vials of hydra to naturalists across Europe. The planimal was a sensation.

Clearly, investigations of minibeasts have the potential to raise fundamental concepts and sustain interest. As a seventeen-year-old student at Edinburgh University, Charles Darwin presented a scientific paper to the Plinian Society that described the free-swimming larvae of *Flustra,* a polyp he had collected along the Scottish coast. Darwin demonstrated that these were not motile eggs but hatchlings.[16] As a polyp, *Flustra* held uncertain affinity to animals. Its seaweed appearance disguised its animal nature.

Throughout the voyage of the *Beagle*, Darwin continued to collect and puzzle over the anatomy of these minibeasts, wondering about their resemblance to the common ancestor of both plants and animals. *Flustra* is now classified as a bryozoan or "moss animal"—its formal name maintains the eighteenth-century ambiguity expressed by two small children upon their first observation of the freshwater polyp, hydra.

The lowly earthworm, another minibeast, intrigued the elderly Charles Darwin. He devoted his last researches to the study of worms and their action on vegetable mold (leaf litter). Whimsical, even childlike experimentation, governed Darwin's study of worms. He determined that his backyard earthworms paid no attention to the "deepest and loudest tones of a bassoon" played by his son, Francis.

"Shrill notes from a metal whistle" also interested Darwin's worms not at all. Nor did wife Emma's piano music or shouts, unless, that is, the shouts were so close that the worm could feel the person's breath or the vibrations of the piano strings. Worms on a table ignored piano music from across the room. They were difficult to irritate with Mozart. However, worms in a pot of soil perched on the piano "instantly retreated into their burrows" when "the note C in the bass clef was struck." Striking "G above the line in the treble clef" had the same effect, he noted.[17]

Darwin was quite precise in his observations, and he made worm science the family business. Fortunately for the worms, heavy metal music did not exist in Darwin's Victorian England.

Darwin proved himself to be a great worm and polyp scientist a century before Piaget occupied himself with the behavior and adaptations of Swiss snails. Piaget's ideas about reasoning inform the design of inquiry and amplify the logical structure of children's investigations—inquiries that ought to resemble Darwin's work in many ways.

Science 5/13's *Minibeasts* curriculum clearly demonstrates the potential for tiny and humble forms of life to fascinate children and the child within the adult. Anatomical, behavioral, and ecological phenomena demonstrated by minibeasts are noteworthy, and minibeast study connects with important turning points in the history of science.

LEGACIES REDEEMED

Piaget and Dewey focused their attention on cognition and learning rather than historical epochs. Yet the biogenetic law echoed through their work. For Dewey, it played out in each instance of inquiry learning. In keeping with the metaphor of mind-as-organism, Piaget substituted the progress of logical reasoning for historical epochs and found a ladder of logic for intellectual development to climb. Dewey understood learning through the lens of original inquiry; Piaget, through an analogy with the evolution of snails. The imagery of recapitulation—the biogenetic law—guided their theorizing.

Logical reasoning and scientific methodology stand as twin pillars of science education. Combined, they constitute the scientific habits of mind aimed squarely at unifying the disciplines. They are the legacy of John Dewey and Jean Piaget and the indirect heritage of the biogenetic law.

Few challenge the value of this legacy. However, John Dewey's emphasis on inquiry learning and Jean Piaget's theory of intellect have conspired to reinforce the myth of the scientific method and the quest for unity. Dewey's legacy includes the five-step oversimplification of scientific methodology and belief that scientific habits of mind have nearly universal application. Piaget tended to dismiss the importance of specific subject knowledge to the growth of a universal form of human intelligence.

Their legacies merit tempering and reevaluation. For nearly three hundred years, children's innate capacity to act upon curiosity has captured the attention of philosophical thinkers. The legacies of Piaget and Dewey respect this wisdom. Spontaneous interest emerges and reemerges in their work as a valued starting point of lessons. Commitment to teaching that values interaction, engagement, and interest follow from their perspectives.

Over sixteen hundred years ago, Saint Augustine expressed a very similar idea. His thought remains pertinent: "There is no doubt, then, that a free curiosity has more force in our learning . . . than a frightful enforcement. Yet this enforcement restrains the rovings of that freedom."[18]

Piaget and Dewey respected free exploration, but knew that its rovings needed restraint in order to develop intelligence or master inquiry. Interest is basic. So, too, is discipline. Augustine did not dismiss "frightful enforcement." He appreciated the need for disciplined thinking.

In Dewey, recapitulation imagery intersected with a conception of democratic living. Inquiry required problems, and the living community supplied them in abundance. Work with community partners placed inquiry within a context of social value, not just one of individual cognitive development. Problems are diverse and evolving; so, too, are disciplines.

Two fundamental questions entertained by Piaget and Dewey endure: first, how to represent the authentic practices, processes, concepts, and experiences of scientific researchers in a child-centered fashion; second, how to frame student inquiry within a context of value to democratic living. Next Generation Science Standards (NGSS) invoke the importance of scientific practices and laudably stress the value of personal interest. However, aligning curricula with abstract crosscutting concepts and generic scientific practices contributes very little to finding tangible answers to the two fundamental questions. Answering them requires bridge-building between personal interest and disciplinary agendas as well as tailoring disciplinary agendas to achieve social value. A universal solution does not exist.

Respecting democratic living as a context for curriculum and honoring personal interest as a starting point for disciplined study lead in many directions. Examples abound of children avidly engaged with the natural world and adolescents partnered with experts to improve their communities. Such diversity within schooling is very hard for society to tolerate, but necessary for student engagement.

In one setting, snail lessons exemplify the productive exploitation of spontaneous interest. In another time and place, kayak construction does. Investigating minibeasts, just one topic among many to choose from, stimulates intellectual growth.

Green building design at the School of Sustainability and leatherback turtle conservation at the Estación Las Tortugas underscore the place of social value in curriculum development. Both projects amplify personal interests. In an important sense, these examples represent the achievement of a crucial standard: connecting students to their local landscapes and communities. In the hands of competent teachers—or in the creative design of curriculum materials—a host of different starting points may lead to a manageable number of important ends, such as appreciating the disciplinary resources needed to solve problems of both personal interest and social value.

NOTES

1. Stephen Jay Gould, *Ontogeny and Phylogeny* (Cambridge, MA: The Belknap Press of Harvard University Press, 1977).

2. Herbert M. Kliebard, *The Struggle for the American Curriculum, 1893–1958* (Boston: Routledge & Kegan Paul, 1986).

3. John Dewey, *The School and Society; The Child and the Curriculum* (Chicago: University of Chicago Press, [1902] 1990): 202–4.

4. Deborah Seltzer-Kelly, "Deweyan Darwinism for the Twenty-first Century: Toward an Educational Method for Crucial Democratic Engagement in the Era of the Institute of Education Sciences," *Educational Theory* 58, no. 3 (2008): 294.

5. http://blogs.slane.k12.or.us/kennedy/students/boat-building/ (accessed February 8, 2015).

6. Kip Ault, "The Importance of Conserving Turtle Eggs: Working with EcoTeach and Proyecto Campanario on Conservation Education in Costa Rica," *The Oregon Science Teacher* 45 (2005): 26–29.

7. Jean Piaget, "Genetic Epistemology," *Columbia Forum* 12, no. 4 (1969): 4. Quoted in Stephen Jay Gould, *Ontogeny and Phylogeny,* 144–45.

8. Ernst von Glaserfeld, "Homage to Jean Piaget," Oikos.org, http://www.oikos.org/piagethom.htm (accessed April 16, 2013).

9. Fernando Vidal, *Piaget Before Piaget* (Cambridge, MA: Harvard University Press, 1994), 40. Vidal continues, "The correlation between morphological adaptation and environmental features thus explained geographic distribution and variability, and constituted a trace of the formation and evolution of species."

10. Jean Piaget, *Science of Education and the Psychology of the Child,* translated by Derek Coltman (New York: Viking Press, [1969] 1971): 153. Originally published in French as *Psychologie et Pedagogie.*

11. Ibid., 152.

12. Sheila Parker, *Science 5/13: Minibeasts* (Hemel Hempstead, UK: Macdonald & Co., 1973; eighth impression, 1989).

13. Anna Botsford Comstock, *The Handbook of Nature Study* (Ithaca, NY: Comstock Publishing Associates/Cornell University Press, 1911; revised, 1939; fifth printing of the revised edition with a new foreword by Verne N. Rockcastle, 1996).

14. Robin Burnett, *The Pillbug Project* (Washington, DC: National Science Teachers Association, 1992).

15. Rebecca Stott, *Darwin's Ghosts* (New York: Spiegel & Grau, 2012): 88.

16. Janet Browne, *Charles Darwin Voyaging: Volume I of a Biography* (New York: Alfred A. Knopf, 1995; repr. Princeton, NJ: Princeton University Press, 1996).

17. Charles R. Darwin, *The Formation of Vegetable Mould Through the Action of Worms with Observations on Their Habits* (London, UK: John Murray, 1881), 26–27. John van Wyhe, ed., *The Complete Work of Charles Darwin Online,* http://darwin-online.org.uk/EditorialIntroductions/Freeman_VegetableMouldandWorms.html (accessed April 1, 2013).

18. Augustine of Hippo, Saint Aurelius, *Confessions of Saint Augustine* (New Kensington, PA: Whitaker House, 1996): 29.

12

The Promising Idea of Core Progressions

The Next Generation Science Standards (NGSS) feature progressions of ideas that are central to organizing thinking in the traditional content domains of life, physical, and earth and space sciences. Aligning with these progressions presented a challenge to the zoo-archaeology team. Their struggle, an example of the negative side of aligning a curriculum with the unifying dimensions of the NGSS, does not diminish the positive aspects of the core progressions.

Each progression has the potential to frame sets of problems or tasks for instruction and performance assessment. One progression from earth and space science describes how rotational and orbital motions change the appearances and relative positions of objects in the sky. Lunar phase problems fall within this progression, for example. Thus, understanding the phases of the moon provides a representative item from the progression of core ideas in earth science teaching as set forth in the NGSS. Figuring out why the moon changes in appearance in a regular cycle offers a problem task bonanza at many levels of difficulty.

Children's books offer a glimpse into this potential. The inviting imagery of children's literature stands in stark contrast to the rarefied constructs framing inquiry assessment. There are after all, as James Thurber has written, *Many Moons*![1] Three-year-olds, prompted by hearing *Goodnight Moon*[2] many times before going to sleep, begin to wonder why the moon appears in the daytime sky. Children who have ridden in cars at night know that *The Moon Came Too*.[3] In college-level astronomy, a few students learn how to predict solar eclipses, as did a famous *Connecticut Yankee in King Arthur's Court*.[4]

GRASPING AT THE MOON

Understanding of lunar phases progresses from childhood awareness of the position and shape of the moon in the sky to geometrical reasoning based upon an orbital model of motion. Thinking about the movement of celestial objects in general faces difficult challenges of visualization and spatial reasoning. Yet these reasoning skills are required in order to master the orbital model of motion and interpret lunar phases.

The orbital conception of motion accounts for shapes and positions as seen from different vantage points. The observer and the objects observed are all in motion with respect to each other, and the observer may observe from different positions. The conception of orbital motion enriches and extends the conception of frame of reference: change of position with respect to time. Interpreting the changing positions of celestial objects in the observer's sky puts frame-of-reference thinking to work.

The progression may start with—and often returns to—straightforward propositions: the earth rotates about an axis; day and night occur due to this rotation. As the earth turns, the moon comes into view and disappears. The moon is difficult to see, or is not visible at all, in a bright sky. The moon orbits the earth.

The portion of the moon illuminated by the sun and visible from earth changes as the moon orbits the earth. These changes are called "lunar phases," and they comprise a repeated cycle. The amount of moon visible from earth increases ("waxes") from the new moon through the crescent, quarter, gibbous, and full phases of the moon. The visible portion then decreases ("wanes") from the full moon through gibbous, quarter, and crescent to new moon again.

The cycle was very important to Princess Lenore of *Many Moons* fame. She asked her father for the moon, the lovely crescent one. She was not worried about taking the moon from the sky. A new moon, she knew, appeared every month. She enjoyed watching it grow.

The key concepts for improving upon Princess Lenore's understanding of the lunar cycle are rotation, orbit, scale, and phase. The phenomenon of interest is the appearance of a distant, illuminated object in motion to an observer also in motion. The challenge is to match thinking to the demands and obstacles of the problem. One of the obstacles is that the moon is very far away and its motion imperceptible to the observer over a short time. Due to its distance, the crescent moon appears to be no bigger than a thumbnail. The real moon cannot hang as a pendant around a princess's neck unless the princess is quite large.

As a result of the relative positions of the earth, moon, and sun, some portion of the illuminated half of the moon is visible by an observer on earth. A typical challenge in school science is to figure out the relative position of the moon with respect to the earth based on its appearance and location in the sky. If walking down the street, the moon seems to follow. It moves from behind one house to the next. The observer needs to record observations of the moon from a fixed position at regular intervals of time over several weeks to improve upon *The Moon Came Too* level of thinking.

A student can memorize relevant propositions about the lunar cycle but without developing a mental model of orbital motions adequate to answering questions about lunar phases. The transformation of phase and position data using the observer's frame of reference to an orbital model of the relative positions of three bodies is daunting. Drawing is essential for interpreting observations of lunar shape and lunar position with respect to the observer's horizon in terms of orbital motion. However, even when taught well and with drawing and modeling exercises (including body movements), students still tend to confuse eclipses with phase changes and cannot picture lunar phases as seen from multiple locations.

San Francisco Moon

Several graduate students in science education gathered in the lunchroom. A former middle-school earth science teacher was sharing her frustration with teaching about lunar phases. She described a novel problem posed to her class following lessons on the lunar cycle. The problem, she believed, was one that all would solve right away. Their frustration had surprised her, and she shared the problem with her fellow graduate students:

An observer in San Francisco enjoys the view of the moon at dusk with the sun just having set. The moon appears close to the western horizon. That same evening, another person enjoys watching the moon in Santiago, Chile, also at dusk just after sunset. What phase of the moon does the person in San Francisco observe? What phase of the moon does the person in Santiago observe?

None of the graduate students answered with any greater success than the middle schoolers. They could not quickly realize that both observers saw the same phase, a crescent.

The horns of the moon point away from the sun. The moon is the letter *C* to one observer, the same letter backward to the other.

Knowing the right propositions, as did the graduate students, does not equate with being able to solve lunar phase puzzles consistent with an orbital model of motion. It takes practice with problems to get there.

The explanation is as follows: rotation makes the sun appear to set. It's dusk, so the sun has just set. The line of sight to the moon is very close to the line of sight to the sun, now below the horizon. The sun is below and the moon just above the horizon. Therefore, the moon must be on the side of the earth toward the sun, almost in line with the sun. Because the bright side of the moon always faces the sun, only a sliver of the moon's bright side is visible from earth. In other words, when the moon is on the earth's sunny side, observers see a crescent moon.

Moving the line of sight from San Francisco to Santiago does not change the relative position of the earth, moon, and sun. The two cities are in different time zones, so sunset happens in San Francisco first and Santiago later. That's a factor that may provoke some confusion. The moon moves in its orbit constantly, but not by much during the couple of hours separating sunsets in the two time zones. Solving the problem may safely assume that neither the moon nor the earth has moved appreciably. Both moon-watchers see the same moon in the same position and in the same crescent phase.

Solving additional problems in terms of orbital motion and the two observers' locations builds upon this first solution: Will the moon set soon or rise higher in the sky as the night progresses? Will the moon wax or wane in the next few nights? And what's up with the crescent's letter *C* reversal?

Keep in mind that a progression must progress to somewhere. Having a good grasp of the destination—for example, interpreting the night sky in terms of an orbital model of motion—sets the destination. The emerging conception guides solving the problem at the same time as struggling with the problem brings the conception into focus. In other words, working on solving lunar phase problems constructs the orbital conception of the earth and moon in space. Good instruction balances challenge and frustration as levels of understanding progress.

"Rote" refers to memorization of propositions: "the moon changes phase; the moon orbits the earth approximately once a month." Students can say these things without knowing their meaning—without being able to solve lunar phase problems as they progress in difficulty. The progression in problems and methods of their solution are what core progressions in the NGSS promise to make plain to teachers. Disciplinary core ideas are more than propositions. They are the tools of thinking and performance is the proper measure of their grasp.

The orbital model of motion holds the key to answering many questions, and many questions may contribute to its construction. Again, it's OK to pay close attention to key propositions whose meaning is not at first apparent. They are often descriptive statements that cannot be obtained by experience.

For example, looking down from above the north pole of the earth, the earth spins counterclockwise from west to east. From this same position high above the north pole, the moon moves counterclockwise in its orbit around the earth. The earth, viewed from above the north pole, also orbits around the sun counterclockwise. The propositions about the rotational sense of the earth are fine starting points, but they need representation as drawings or manipulations in order to stick.

The crescent moon has been moving counterclockwise from a position directly between and in line with the earth and sun to its position as seen at dusk from San Francisco and Santiago. As a consequence of this movement the moon changed from its new moon phase to a crescent phase. Only a sliver of the illuminated half of the moon can be seen by an earthbound observer. As it continues to move away from the direction of the sun and to the "side" of the earth, the illuminated portion visible from earth increases, or waxes.

The earth must turn farther each night in order for the western horizon to block an observer's view of the moon. First the sun sets, and then the moon sets, and that happens later each night as the phases progress through the lunar cycle. At sunset in successive nights, the moon hangs higher and higher in the western sky, waxing its way to quarter status.

The horns of the moon point away from the sun. In effect, the San Francisco observer is upside down with respect to the observer's counterpart in Santiago. The Chilean sees a crescent that looks like the letter *C*. The American sees it reversed and inverted. The *C*s are tilted, not vertical, and for a tropical observer, the figure formed is a bowl.

Lunar *C*s often trouble comic strip artists. In the northern hemisphere, the backward moon sets shortly after the sun. A proper letter *C* precedes sunrise in the early dawn and is the last glow of the waning moon. Watch movie sets and comic strips to see if the artist got it right or wrong from a northern point of view. Of course, drawing crescent moons from a Chilean perspective is perfectly acceptable.

The reason for the letter reversal is no more difficult than describing the direction of flow of a stream. Stand on one shore and the stream flows from left to right. Cross the stream, face it, and it flows from right to left. The stream did not change course; the observer rotated 180 degrees. A Santiago observer, from the frame of reference centered on San Francisco, is upside down, another 180-degree flip. These are frame-of-reference problems.

The graduate student seized on the "flip" idea to help her fellow students recognize the solution to the problem. She took a marker and drew a crescent moon setting in the west on a window, then sketched a stick figure person. She held the sketch and asked, "Stick figure person, do the horns of the moon point to your left or to your right?" A stick figure voice (probably one of the graduate students) answered, "To my left."

Next the former middle-school teacher pretended to trot the stick figure around the earth to Santiago. She held it upside down facing the same crescent moon on the window. "Stick figure person, do the horns of the moon point to your left or to your right?" Right away all of the graduate students in science education realized: "The horns point to my right." Interpreting lunar phases sometimes require topsy-turvy thinking.

All were prepared to progress to the next level of difficulty. Suppose on some other night the San Franciscan gazes skyward to witness a stunning crescent, a letter *C* moon in the sky. Where's the moon? What time is it, more or less? What's the phase?

The moon must be in the east, the time near dawn, and the waning moon about to enter its new moon phase. That's the power of a mental model of orbital motion.

The Lunar Objective

The concept of "lunar phases" holds a great deal of potential to structure a progression of questions. "What are the phases of the moon?" is one. "In what order do they change?" is another. They continue on up the ladder toward the objective, "Given a phase of the moon, determine its approximate rising and setting times."

Keeping a moon log of rising and setting times corresponding to phase can precede creating diagrams able to account for these sightings. Manipulating physical models contributes to understanding as well. The point is not that everyone needs to be able to infer lunar phases for people all over the earth through the month. The point is that reasoning about lunar phases and their predictability develops as an internalized mental model of orbital motion. There is dramatic potential for the progression of understanding orbital motion and its consequences for observation from earth, unfettered by alignment with crosscutting concepts and scientific practices. Three-dimensional thinking about the moon in space builds upon the disciplined thinking of an orbital model.

The abstract "dimensions" of the NGSS provide little guidance in generating performance expectations and designing tasks to gauge progress in understanding core ideas. Greater attention to the problem-solving role of core ideas remedies this situation as task difficulty progresses.

Imagine repeating this scenario for the rising and setting times of constellations. All such problems depend upon reconciling a spatial model of orbital motion, including rotation, with the sky as observed in everyday life. Frame of reference is key. One obstacle to the visualization of objects in motion is time scale. Positions of celestial objects change without appearing to move as they are watched.

At what age should a child be expected to master the concept of lunar phases? By age three children are surprised to see the moon in the daytime sky. At age five many believe that the moon goes behind the sun during the day. Solving the Santiago-San Francisco lunar observer problem of the horns being reversed when the moon is in the same phase frustrates most adults. Determining "grade-level proficiency" for the concept of the lunar cycle is a fool's errand. The progression is what matters, with the destination carefully conceived.

Middle schoolers often manipulate orbital models. Somewhere between grades six and twelve, expecting students to be able to use a mental model of orbital motion represented by two-dimensional drawings or manipulated objects seems reasonable. Just knowing the right answers in propositional form falls far short of this target at every stage. Questions about lunar phases progress in difficulty from recognizing a phase, knowing the sequence of phases, explaining the moon's changing appearance in terms of lines of sight (and not the earth's shadow), and predicting rising and setting times at different locations and for different phases.

Many students confuse the phenomena of lunar eclipses and phases. It's a common misconception. During an eclipse of the moon, the passing shadow of the earth maintains a curved edge. There's no moment when a straight line bisects the moon, as occurs when the moon is at first and last quarter phases. That observation has the potential to dispel a misconception. Teaching in terms of progressions should acknowledge likely misconceptions—psychological obstacles and confusions that commonly trouble learners. The right model does wonders.

When Princess Lenore grew older and had improved upon her understanding of the lunar cycle, she asked her teacher for a scale model of the earth-moon system to keep in her closet. "That's easy," said the teacher who had recently attended the Harvard-Smithsonian Center for Astrophysics's *Private Universe Project* workshop and also owned a personal copy of ESS's *Daytime Astronomy*. "The earth's diameter is about four times the moon's, and both are spheres. Let's get a ping-pong ball and a plastic bead one-fourth its diameter, about one centimeter." The teacher glued the ping-pong ball to one end of a stick and the bead to the other. She used a stick 110 cms. in length, the approximate number of moon diameters separating the earth from the moon.

The earth-and-moon-on-a-stick delighted Princess Lenore. "One more question," she added. "What will the path of the moon look like with respect to the sun through a year? I know now that the earth-moon system itself orbits the sun. I wonder: Does the moon's path snake back and forth or do loopy-loops?"

"That's not so easy to figure out," answered her teacher. "We can try drawing the moon's path around the sun to scale. Instead of 110 cms. as the lunar orbit radius, we'll need something much smaller."

A quarter worked well. With the earth imagined as a point at the center, the circumference of a quarter may represent the orbit of the moon. At this scale, the earth-moon system (the quarter) orbits several feet from the sun in a large, circular arc. The quarter—earth and moon together—moves along this arc surprisingly far each month. Every week the moon circles one-quarter of the quarter as it advances, changing from new, to first quarter, to full, then third quarter, and back to new moon again. At this scale, the earth and moon are little dots.

They moved the earth ahead along the arc drawn of its orbit one month at a time. "Look, look," said Princess Lenore. Words from her childhood popped into her mind. "The moon came too!" It did and it traced a path that never crossed itself. The path curved, always slightly concave toward this sun, a shape similar to the interior combustion chamber of a Wankel rotary engine. "Oh," the Princess thought, "I think it could be a cyclically closed pericycloid!"[5]

CELESTIAL OBSERVATIONS

Sometimes misconceptions about the earth in space, the phases of the moon, and the reasons for seasons prove tenacious. Even students who readily accept a cosmology where spherical bodies orbit each other may harbor a sense that space itself has a top and bottom, hence an up and a down in an absolute sense.[6,7] A conception of absolute space existing independent of the bodies that occupy it is certainly not silly or naïve. It undergirded Newton's worldview.

The orbital conception of motion, with consequences for observers' interpretations of celestial phenomena based on their choice of a frame of reference, develops across a decade of science teaching. Its grasp is irreducible to right-answer propositions. Instead, success with problem-solving exercises displays progress toward its mastery.

J. Myron Atkin's *Charting the Universe*[8] introduced such a curriculum in the 1960s. His program taught how lines of sight from the endpoints of a long baseline converge at the moon to form a long, skinny triangle. By drawing a similar triangle in direct proportion to the one with the moon at the apex, lunar distance can be determined. *Project STAR* (Science Teaching from its Astronomical Roots)[9] continued in the tradition of Atkin's materials.

Misconceptions

Misconceptions about the earth's geometry and orbital motion have their origins in everyday perspective.[10] Scale is an obstacle, and techniques are needed to overcome the biases of everyday perspective, from the sense that the earth is flat to the belief that the stars occupy fixed positions. In addition to this difficulty, some of the tools of instruction introduce unintended and unanticipated misconceptions. For example, think of a globe.

Often drawn on the Pacific Ocean is a figure-eight form labeled with the months of the year. People may misconstrue this figure as somehow being a representation of earth's motion. The diagram—an analemma, to be precise—graphs the difference in time between the mean solar day and the apparent solar day. It also depicts the departure of the sun vertically and horizontally from its mean position in the sky day by day.

The definition of an analemma seems abstract and complex. It is mathematical. However, the 1960s earth and space science module *Daytime Astronomy*[11] introduced upper-elementary grade students to a host of activities for charting the position of the sun through the calendar year. Students collected data based on changes in the lengths and bearings of the shadows of a stick (or "gnomon"). The exercise formed the basis for interpreting the apparent movement of the sun from the earth's rotation and revolution, including the inclination of its rotational axis. Those classes that carefully recorded shadow tips at precisely twenty-four-hour intervals for a year (missing some days was OK), noticed a pattern when they connected the dots: a figure-eight shape, or analemma—the apparent early or late, high in the sky or low in the sky, return of the sun at equal intervals of clock time.

The 1983 documentary *A Private Universe* from the Harvard-Smithsonian Center for Astrophysics,[12] filmed one of the best students in an eighth-grade earth science class drawing the shape of the earth's orbit. Her earth orbited the sun, but made two strange loops at opposite ends of its orbit. The looping pattern suspiciously resembled an analemma, the figure eight often printed on a globe.

Monthly segments mark off the analemma as a means of adjustment to sundial time. Many cursory looks at a Pacific Ocean, looping figure labeled like a calendar may have triggered the strange drawing of loops at the opposite ends of the earth's orbit. The film did not pick up on this point. It did emphasize how often students, even Harvard graduates, accounted for seasonal change in terms of the changing distance between the earth and sun—a persistent, common misconception.

In summary, the methods for investigating celestial motions must respond to the obstacle of scale and overcome the biases of everyday perspective. Through drawings and problem-solving tasks accompanied by verbal explanations, students may express a wide range of understandings of the earth in space. Their responses often reveal idiosyncratic as well as widely shared and persistent misconceptions.

Up Up, North North

Ask a group of teachers to go outside with a bunch of globes and pair up. Challenge each pair to position the globe so that the sun illuminates the globe the same as the earth at that very moment,[13] and then write down the directions for doing so in as few steps as possible. Right away they call upon presumably relevant knowledge: the time of day and the season of the year. Some may try to rotate the earth as if moving the hands on a clock. Most will talk about "tilting" it on its axis with respect to the sun.

Several sets of directions do the trick. The simplest set is: "Wherever you are, hold the globe so that 'up' on the globe is up to the sky and 'north' on the globe points to true north. Just make up, up and north, north." To realize why this works, imagine a tiny globe person standing on the globe in exactly the same place as one of the teachers. "Point up!" The tiny globe person and the real earth person, if standing in the same relative positions with respect to the sun, both point in the same direction. Globe up and earth up are aligned when they do.

Another way to give this direction is, "Hold the globe with your location at the top." That aligns the two ups. Everyone looks at the sky as if standing on top of the world.

Now, point north. Finding north may present a problem. One way is to mark shadows for a while during a midday hour or so. The shortest shadow, cast when the sun is highest in the sky, must point north. Making up, up and north, north—with the observer's location on top of the world—aligns the globe in the sunlight the same as the earth. The procedure works for anyone, anywhere, and anytime.

The shadow on the globe then corresponds to where night has fallen on the earth. Little plastic people with "clay feet" (literally) may be moved around the globe. They will cast shadows similar to those cast by real people at the same places on the real earth. Where the sun is directly overhead at that moment, a little plastic person will cast no shadow. Little plastic people with long shadows stand where the sun appears to rise and set. The pop-up book *Anno's Sundial* builds upon this geometry.[14]

A globe oriented the same as the earth in natural sunlight has great potential for working through concepts about the relative position of the earth in space and the location of an observer on earth. It's a method matched to the demands of the problem. It's also an excellent example of a task amenable to performance assessment, an excellent practice encouraged by the NGSS.

Butterflies and Arrows

The progression of understanding through disciplinary core concepts is seldom simple and straight. As Michael R. Cohen, Professor Emeritus of Science Education at Indiana University/Purdue University–Indianapolis, often reminded his students, "Learning is more like the flight of a butterfly than the path of an arrow."

There are confusions to overcome and intuitions at odds with accepted explanations to reconcile. Students need to learn how to navigate through these confusions toward the interpretation of both everyday experience and carefully crafted observations in terms of a well-validated conception. A good conception organizes thinking in a well-defined context. There are not too many of them, and they function to assemble innumerable, discrete pieces of information into a coherent whole—the orbital model of motion, for instance.

These conceptions, and their relationships to methods of inquiry capable of overcoming obstacles inherent in the nature of different phenomena, are the proper target of learning in science, not grand, unifying abstractions. The NGSS do harbor these conceptions, and they merit amplification more than alignment. They get moving with MS-ESS1-1: "Develop and use a model of the Earth-sun-moon system to describe the cyclic patterns of lunar phases, eclipses of the sun and moon, and seasons."[15]

NOTES

1. James Thurber, *Many Moons*, illustrated by Louis Slobodkin (New York: Harcourt, Brace & World, 1943).

2. Margaret Wise Brown, *Goodnight Moon,* illustrated by Clement Hurd (New York: Harper Collins, 2005).

3. Nancy White Carlstrom, *The Moon Came Too*, illustrated by Stella Ormai (New York: Macmillan, 1987).

4. Mark Twain, *A Connecticut Yankee in King Arthur's Court* (New York: Oxford University Press, 1996).

5. Piotr Drogosz, "Geometry of the Wankel Rotary Engine," *Journal of KONES Powertrain and Transport* 17 (2010), http://ilot.edu.pl/kones/2010/3_2010/2010_drogosz_geometry_of_the_wankel.pdf (accessed February 14, 2015).

6. Joseph Nussbaum, "Children's Conceptions of the Earth as a Cosmic Body: A Cross Age Study," *Science Education* 63 (1979): 83–93.

7. Cary I. Sneider, and Mark M. Ohadi, "Unraveling Students' Misconceptions about the Earth's Shape and Gravity," *Science Education* 82 (1998): 265–84.

8. J. Myron Atkin, and Stanley P. Wyatt, *Charting the Universe: Elementary-School Science Project* (Urbana, IL: The University of Illinois, 1961).

9. Harvard-Smithsonian Center for Astrophysics, *Project STAR: The Universe in Your Hands* (Dubuque, IA: Kendall Hunt, 2001; Cambridge, MA: Harvard-Smithsonian Center for Astrophysics, 1990).

10. Charles R. Ault, Jr., "The Everyday Perspective and Exceedingly Unobvious Meaning," *Journal of Geological Education* 32, no. 2 (1983): 89–91.

11. Norma Arrow, Edith H. E. Churchill, and Rose Lea Crowley, *Elementary Science Study (ESS) Daytime Astronomy Teacher's Guide* (New York: Webster Division, McGraw-Hill, 1971).

12. Matthew H. Schneps, and Lindsay Crouse, *A Private Universe* (San Francisco: Astronomical Society of the Pacific, 1989): video.

13. Victor E. Schmidt, and Verne R. Rockcastle, "Sunlit Earth," *Teaching Science with Everyday Things* (New York: McGraw-Hill, 1982).

14. Mitsumasa Anno, *Anno's Sundial* (New York: Philomel Books, 1987).

15. NGSS Lead States, "MS-ESS1-1: Earth's Place in the Universe," in *Next Generation Science Standards: For States, by States* (Washington, DC: National Academies Press, 2013), http://www.nextgenscience.org/ms-ess1-1-earths-place-universe (accessed February 10, 2015).

13

The Elbow and Ankle Tour of the Zoo

The relative positions of joints that assemble a skeleton and articulate its limbs suggest evolutionary affinities. The more similar the limbs, the more similar the kin, up to a point. That's a starting point for using the concept of "homology" as a tool of inquiry. Thinking in terms of homology means reasoning in Darwinian fashion and the "second style" of science. As in the case of the earth in space, concept and method reinforce each other. There is no need to detour onto the highway of big ideas common to all the sciences in order to make progress. Core progressions suffice.

Sketching animal limbs from torso to toe-tip engages zoo visitors in comparative anatomy and focuses attention on the positions of ankles, elbows, knees, and wrists. At first the young artist, looking at a large quadruped, notes, "The knee bends backward." Another asks, "Do elephants even have knees?" Another adds, "Looks to me like the stork's knee bends backward, too."

THE LIMBS OF *LOONEY TUNES*

Knowledge of comparative anatomy, especially limb structure, served cartoon animator Chuck Jones, creator of Bugs Bunny, Daffy Duck, Wile E. Coyote, and many other familiar characters, very well. He perfected both form and movement in his animations, obviously anthropomorphizing the faces of his creations, but at the same time keeping most of the limb proportions and limb joints in their proper configurations.[1]

A powerful hop in an unanticipated direction is a good way to escape from a predator. Bugs Bunny embellishes his escapes with numerous tricks, to the consternation of Elmer Fudd, but he still runs on proper bunny feet. The speedy Roadrunner, abetted by similar trickster stunts, forever seems to frustrate Wile E. Coyote. Roadrunner and Wile E. Coyote have human attributes, yet both have limbs in proper bird and coyote proportions.

In his autobiography, *Chuck Amuck*, Chuck Jones recounts the story of animating Rudyard Kipling's *The White Seal* as a Disney production. Forgive the cartoon animator's failure to distinguish between seal and sea lion for a moment. What the Disney people expected was animation that captured pinnipeds (seals, sea lions, and walruses) swimming in a convincing manner. Sea lions do have several traits that distinguish them from seals, such as small, external ears. However, sea lion swimming parallels seal swimming because both utilize similarly structured flippered or "finned" feet, modified appendages descended from their land-living ancestors.

Pinniped Grandchildren

Jones understood the relationship between structure and function. He realized that anatomy causes animals to move as they do, and their motions correspond to their way of getting about in a particular environment. To improve his animation, Jones went to the San Diego Zoo, where he and his fellow animators studiously watched sea lions:

> I could not conduct a course in comparative anatomy with my animators. I was not educationally equipped to do so; therefore, I followed the most logical substitute—comparing my own anatomy to that of the sea lion. This is not as difficult as it might sound, since all vertebrates have more structural matters in common than differences. Our bones and muscles all bear pretty much the same names and are readily identifiable; the great differences are primarily in length and weight of the bones and the musculature, and, of course, in the skull structure.[2]

Jones concluded that the sea lion's flippers were hands and feet, not fins, because they had obvious toenails and fingernails. He guessed that the upper arm bone (humerus) essentially remained within the body and that only the external arm from the elbow joint outward emerged. Sea lion bodies hid sea lion elbows. How did Jones test this hypothesis without conducting a sea lion dissection?

Back at home and away from the zoo, he recruited his two grandsons, ages eleven and thirteen. Next he tied their arms to their bodies from the shoulder to the elbow and their legs together as well from hip to ankle. He helped them don swim fins on both their hands and their feet. Boys with arms literally pinned to their bodies and legs bound together, thought Jones, would behave themselves like pinniped mammals.

With Jones's colleagues assembled as an audience, into the pool tumbled the two highly animated grandsons. They had only one way to swim: the same as a sea lion. They executed sweeping dives, though with a tad less grace than true sea lions, and surfaced to breathe in a great hurry.

Yes, this was a loony idea. Do not try this experiment at home with your own children, grandchildren, or other people's children, for that matter. In fact, it would be a bad idea to try this experiment on yourself, especially with no lifeguard on duty. The good news: the boys survived just fine, although they spent the rest of the afternoon bouncing beach balls on their noses.

Shoeing the Horse

In *Chuck Amuck*, Jones's sketches of Elmer, Daffy, Bugs, and Porky Pig appear on the same page as his sketches of horse, human, dog, cat, and cow limb. The caption states, "Same bones—different lengths only."[3] Jones the animator is also Jones the homologist.

Jones superimposed the body forms of the cartoon characters on very basic lines or axes. The lines corresponded to curves, angles, and orientation of the backbone and limbs positioned for motion. Change the angle, change the length, change the character, and change the action.

Humorous, yes, but more importantly Jones's cartoons teach how to invent new, strange, and silly creatures by varying the proportions, angles, and basic shapes used to draw real animals quickly. His techniques keep these creations believable as skunks, ostriches, or roosters.

With just modest practice, even novice artists may succeed in rendering realistic drawings of zoo creatures. A good training exercise in this regard is to play Nike designer to the tetrapod kingdom. Jones challenged his animators to design a pair of sneakers that would properly fit a horse, a kangaroo, or a sloth—Nikes for every species.

Recognizing how and where the joints of the foot and leg bend are essential to this task. Think: length, angle, action. Which parts of the foot, ankle, and shin belong inside the sneaker, and which parts do not? That's the key. And sneakers, of course, never cover knees.

As pointed out by Jones, on a human, a sneaker covers the ankle. A Nike designer will want kangaroo, horse, and sloth sneakers to do the same.[4] If a kangaroo were to play basketball while wearing the proper footgear, the shoe, as Jones realized, must tie just above the ankle. Where's the ankle on a kangaroo? It's the back part of the foot that touches the ground when the kangaroo is at rest. That is precisely the same part of the human foot that touches the ground when at rest. A standard Nike basketball shoe, size 100, would be a good approximation of a sneaker fit for a kangaroo.

Socks look sharp when pulled up to the knee, placing the venerable swoosh in clear view. So on which part of a kangaroo leg do you draw the sock? The kangaroo's knee joint lines up with its belly. It's going to need an extra-long pair of tube socks.

The challenge of fitting a sneaker to a three-toed sloth or a draft horse (the Percheron in *Chuck Amuck*) is somewhat greater. In the sloth, the joints are more subtly hidden by the blobby body, but the comparative strategy suggests, joint by joint, where to look. The sloth sneaker will most definitely require strongly reinforced toes, given its long, curling claws.

The knee on the horse caps the base of the femur of its hind leg. The ball end of the femur fits in the socket of the hip. For the most part, the femur (thighbone) runs from belly height and above; the horse's knee thus placed well above the ground. The leg bones swing at the hip, bend at the knee, and angle again at the ankle joint. The knee joint points forward and the ankle joint toward the horse's posterior and upward. The horse's ankle sits rather high as well. When shoeing a horse with a human shoe, the sole runs from toe to just beneath the ankle. Much of the sole never makes contact with the ground. The horse actually perches on its toe and the single toenail at the end of each foot thickly expressed as a hoof. Maybe that's how the "*Perch*eron" got its name.

Humans play basketball quite well with just one pair of shoes and naked hands. Because they have legs and not arms, horses need two pairs. To arrange a fit similar to the hind limb, the forelimb shoe must enclose the knuckles (or "fetlock" on the horse) and reach to just above the horse's wrist (or "knee") partway up the front leg. Horses can neither dribble nor shoot well with their forelimbs. However, dressed in properly fitting front-foot sneakers, they can set awesome picks.

Elephant Knees and Giraffe Elbows

Imagine touring a zoo guided by Chuck Jones and his animator team. Enter with an artist's sketchbook in hand, asking, "How do limbs differ from creature to creature?" The task is to sketch the limbs of a variety of creatures from their shoulders and hips to the tips of their digits on the fore and hind limbs.

Visitors may notice that the hind leg of an elephant appears to bend in the wrong way—zebras, too. That's because of the elevated ankle and toe-standing stance. The ankle on many grazing creatures seems to be where the knee belongs. Do elephants even have knees? The long forelimb of the giraffe prompts an analogous question, "Do giraffes have elbows?" In both cases the answer is yes, but convincing oneself that this is so requires tracing the bones from joint to joint, noting the angles as they flex, and placing the segments in one-to-one correspondence with one's own limbs. Homology cues the thinking.

The elephant's toe-standing stance may prove difficult to discern, and counting its toes is somewhat confusing as well because of the fatty pad in the back of the foot. How strange to think of a jumping kangaroo sitting back on its ankles while the bulky elephant tiptoes about.

Jones realized that the hip of a sea lion was located near the end of the body where the back flippers flared out. Of course, the human torso has a hip girdle at its lower end as well, but the legs put the human hip about midway between feet and head. In the sea lion, feet are almost at the hip. This was the arrangement mimicked by binding the two boys' legs together so that they could only freely move their feet, like a seal or sea lion.

SKETCHY LOOKING

Sketching holds captive in the imagination the great diversity of limbed life. So distinctive are the feet and toes of hippos and storks, rhinos and bears, zebras and sea lions! Within these extreme differences deep similarities remain. The forelimbs of naked mole rats and fruit-eating bats end in fingers—clawed in one case, webbed in the other. The bone lengths vary, but the sequence of joints align from species to species, exactly as taught by Chuck Jones, Charles Darwin, and Georges Cuvier, comparative anatomy's founding father.

On a brisk fall day, a class of prospective teachers visits the zoo. Lead by an art and science educator duo, they are there to learn to sketch limbs in the manner of Disney animators. Comparisons among limbs introduce them to the concept of homology. Limb anatomy also suggests adaptations to habitat. Limb structures match limb functions—running, digging, and climbing. Limb homologies indicate the degree of relationship to a common ancestor. Sketches of limbs help to answer questions such as, "Are rhinos more closely related to elephants than they are to horses?" and "What good is a hoof?" Touring the zoo will never be the same again after participating in this exercise. Nor will watching pets.

Pencil in hand, the amateur artist's ankle and elbow tour of the zoo begins. It leads to an understanding of basic principles of evolution and ecology at an elementary level.

Careful portraits of lumbering elephants and tottering penguins, posed in a series of positions, remains for now the professional animator's dream. Achieving realism, elegance, and complexity is not the goal: the line's the thing. In quick sketches, line segments depict angles and lengths. From shoulder to toe, the segments zig, then zag, some short, some long. The back may arch, run straight, or sag. With training, getting the proportions right becomes possible. A thumb moves along a pencil held in the line of sight, the pencil's angle and the segments marked by the thumb's position transfer to paper in correct proportion and orientation.

Stick rhinos and stick hippos appear. The elephant's massive body reduces to skinny lines and angles. The necks of storks and giraffes seem almost silly in proportion to their bodies.

Trapezoids, triangles, squares, circles, ovals, and rectangles create heads, necks, limbs, haunches, shoulders, ears, muzzles, hooves, and tails. Soon a stocky mountain goat, then a bulbous warty hog, later a rotund orangutan, and eventually a long-limbed gibbon peer back from the sketchbook.

Two steps remain: (1) contouring the outline of the animal's profile and (2) finishing with shading. The process is Picasso in reverse: contouring the blocky impressions yields more realistic forms. Adding details of shading plus impressions of feathers or fur completes the entry.

The ankle and elbow tour of the zoo guides attention to the proportions and shape of the body. By going out on a limb to its nailed, hoofed, padded, or clawed end, the artist hypothesizes the relative positions of knees, ankles, and elbows. Each change in angle implies a joint, shoulder to foot. The sketching mantra does the trick: make a stick figure, block in basic shapes, contour the edges, and add some shading.

Fine-tuning the digits completes the task. Do claws, hooves, or nails come in singles, doubles, triples, or more? Do they turn inward or outward or backward or align straight ahead? Which way do the hind toes point on creatures that climb: the sun bear, the margay cat, the local squirrel that chanced to run by? Note which parts of the foot or hand touch the ground when the creature walks and whether these contact points are the same or different when it stands or runs.

Attention to drawing details slows down the all-too-often frenetic tour of the zoo. Style of movement, adaptation to habitat, and degree of common ancestry: all are bound up in limb structure. The same could be said of skulls. Comparing limbs by means of sketches prompts archetypal Darwinian questions: "How closely related are dogs and cats? Dogs and bears?" Only the cats, for example, can retract their claws. Claws are always visible on bears and dogs. One trait is not definitive, but the retractable claw does suggest that dogs and bears are the closer kin, with cats more distant on their tree of common ancestry.

HOMOLOGY AS METHOD

Darwinian histories portray the stunning creativity of natural selection. The struggle for existence over time turned lobe-finned swimmers into walkers, walkers into runners, and then some into flyers and back into swimmers. Limbs that changed through tens of millions of years of evolution have left telltale signatures in the joints.

Among a long-standing group of terrestrial mammals, at the end of the leg and start of the foot sits a short ankle bone, rounded at both ends like a double pulley: the astragalus. This bone defines the artiodactyls: antelope, goats, pigs, cows, sheep, and deer. These are even-toed, hoofed (and quasi-hoofed), toe-perching runners. Whales may be one of their offshoots. Horses, donkeys, rhinos, and zebras do not belong to this group of even-toed creatures; their toes come in odd numbers.

Remarkably, an extinct, limbed whale, *Ambulocetus*, sported an astragalus bone at its ankle joint—in effect, at its heel. Strange cousins emerge from this curious anatomical fact. "Is a whale closely related to a moose and its artiodactyl brethren?" Probably so, but likely more so to hippos.[5] Somewhere, someday—probably in Pakistan—a group of fourth graders may find themselves picnicking next to a set of tracks left by a walking whale.[6]

Classifying birds on the basis of limb anatomy ties birds to dinosaurs.[7,8] Wrist bone homologies help to solve the bird origins puzzle; similarities among ankle bones, the whale one. These details are not possible to sketch in a zoo tour; skeletal samples of extant and extinct species are necessary. Nevertheless, the conception on which these inferences rely remains stable. Its usefulness reaches widely.

The children of *Sesame Street* sing, "One of these things is not like the other, one of these things just doesn't belong." Horses trot to a different song than cows, goats, and antelopes; the horse just doesn't belong. The ankles and toes tell why. The horse's hoof is a single toe, an ancient odd-number pattern, the same as typified the three-toed horses of the fossil record. The same pattern tracks rhinoceros fossils from extinct gracile to modern robust forms. According to limb structure, horses and rhinos are quite closely related to each other. A good name for their common ancestor would be a "rhinorse" (of course).

Is a rhino more closely related to an elephant or to a zebra? In terms of limb similarities, the rhino-zebra match is striking. Are giraffes related to zebras more or less than they are related to storks? Relating storks to giraffes appears to be a stretch. In terms of hoofiness and hipness, knee jerks and ankle twists, the stork is the one that just doesn't belong. Its wing is its thing.

Limbs are marvelously adapted contraptions. Viewed through the lens of homology, sketches of limbs joint by joint from torso to toe unlock mysteries of their evolution at an accessible, elementary level. Zoo sketching need not extend to animating, but Jones's primary insight is telling: structure determines how animals move. As icons of evolution and ecology, limbs do explanatory double duty.

"Homology" is a quintessential tool of inquiry and the zoo is a good place to learn it. Method and concept intertwine in the ankle and elbow tour. The Next Generation Science Standards (NGSS) build upon the "intertwine" image very deliberately, as well they should. The intertwine functions best, to use a metaphor, when it is part of the structure of a discipline. "Structure and function" certainly keep company with the big ideas thought to unify the sciences. Better to keep this idea in context where it might serve a recognized, tangible purpose such as understanding animal adaptations and inferring evolutionary affinities. That's its core disciplinary idea home, a place where thinking and doing intertwine and zoo sketching promotes the progression of learning.

NOTES

1. Chuck Jones, *Chuck Amuck* (New York: Farrar, Straus and Giroux, 1989).
2. Ibid., 258.
3. Ibid., 260.
4. Chuck Jones tells this story in his chapter, "How to Make a Tennis Shoe for a Percheron." As an Oregonian and Portland Trailblazers fan, I have taken license to use the example of a basketball shoe. Jones does sketch a sneaker that covers the ankle of his drawings of horse, human, kangaroo, and sloth. In his sketches, for all intents and purposes, the shoe is the classic, Converse All Star, canvas basketball shoe.
5. J. G. M. Thewissen, Lisa Noelle Cooper, Mark T. Clementz, Sunil Bajpai, and B. N. Tiwari, "Whales Originated from Aquatic Artiodactyls in the Eocene Epoch of India," *Nature* 450 (2007): 1190–95, doi: 10.1038/nature06343 (accessed October 17, 2011).

6. Philip D. Gingerich, "Evidence for Evolution in the Vertebrate Fossil Record," *Journal of Geological Education* 31, no. 2 (1983): 143.

7. John H. Ostrom, "Archaeopteryx and the Origin of Birds," *Biological Journal of the Linnean Society* 8 (1976): 91–182.

8. Richard O. Prum, and Alan H. Brush, "Which Came First, the Feather or the Bird?" *Scientific American* 288, no. 2 (2003): 84–93.

14

Cracks in Unity's Armor

The Next Generation Science Standards (NGSS) put an end to engineering's second-class citizenship in K–12 education. That is the most obvious crack in unity's armor.

The distinction between historical and experimental styles is widely accepted among historians and philosophers of science. The "second style" opens the door to stressing the importance of plurality and diversity among scientific enterprises. The methodology and rhetoric of the historical style of science has already cracked unity's armor and there's no *a priori* reason to limit the style count to just two. Methods of inquiry typified by the historical style differ from single-manipulated-variable designs with randomized assignment of treatments to controlled or manipulated conditions.

The social value dimensions of science further strain unity's armor. Ethical aspects permeate the protocols of medical science. Climate science and medical science address problems of extreme significance to everyone, but with methods irreducible to stereotypical scientific method. In effect, they add "computational modeling," as do engineering and chemistry, to the styles of science. Other fields invent methods for examining phenomena in natural, not laboratory, settings.

Complex living systems and dynamic earth systems have variables whose interactions are best described in probabilistic ways. Many of their phenomena of interest simply do not happen in laboratory settings; therefore, investigators must work in natural settings. Field studies employ methods suited to field conditions.

Several sciences share vital interests in figuring out complex, natural systems. They address weather, animal behavior, stars and galaxies, earth history, and more. Entangled webs of variables interacting over long periods of time present challenges to the study of such phenomena. Researchers look for descriptive and correlative trends in order to tease out causal relationship from their field data that account for events irreproducible in lab settings.[1]

CHILDREN'S FIELD SCIENCE

Monkeys, for example, behave differently—and perhaps in more demented ways—on monkey islands in zoos than in their preferred habitats. In general, animals display a wide range of behaviors in the wild but with many of their behaviors displayed at low frequencies. In captivity, however, the reverse is true: they display a narrow range of behaviors at high frequency, the ultimate being stereotypical behavior—compulsive pacing and other repetitive acts that worry zoo visitors. Laboratories may even further isolate monkey behavior from the complex natural conditions that trigger it.

Schoolchildren in the United States lack opportunities to observe how primates behave in their native habitats, as did Jane Goodall and Dian Fossey, field study virtuosos. Extensive, descriptive field notes—always attuned to novel and unexpected events—anchored their work. Zoos offer a subsitute to a limited degree, but small creatures native to backyard school habitats offer the greatest opportunities to learn the methods of field study. The field notebook is an essential entry point.

Lizard Studies

The study of common creatures has the potential to align school science with the field practices of contemporary science, claims science educator Mark Windschitl of the University of Washington. He has described "Lizard Studies" by young children as an example of school science closely resembling what field biologists actually do.

In their lizard studies, second-graders kept records of what lizards ate. Fourth-graders modified natural conditions and built an enclosure to find out what lizards did over winter. Windschitl refers to these investigations as "descriptive." Children who inquired about changes in lizard abundance with respect to temperature and rainfall were engaged in "correlative" studies.

Later on, the fourth-graders worked with local farmers to gather data on seasonal lizard movement. This study was "comparative."

Pillbug Projects

Robin Burnett, author of *The Pillbug Project*, proposes, in effect, correlative studies with isopods that engage students both in the field and in the classroom.[2] Isopods, small crustaceans, include two very different terrestrial species, commonly referred to as sowbugs and pillbugs. Sowbugs flatten themselves when in danger; pillbugs roll into little balls the size of peas. As the summer progresses, temperatures rise, and arid days increase, do the relative populations of sowbugs and pillbugs shift? What factors favor pillbugs over sowbugs? The data in answer to such questions range across Windschitl's three categories and can yield causal findings.

Isopods seem to bump into and crawl over each other with abandon. They cluster under logs, rocks, and flowerpots. What do they eat? That question calls for a descriptive study, the same as for lizards. Are they social? Do they communicate?

These questions are much more difficult to answer and depend in large measure on the meaning of *social* and *communication*. Even in field studies, concept and method function together.

As Windschitl and his colleagues acknowledge, experiments permeate biology. Their point is that diverse methods are needed to understand life well, and some of those methods must adapt to natural settings. In some cases, such as building an enclosure for wintering lizards or in setting out a variety of foods to test ant foraging preferences, investigators do manipulate variables and modifiy natural conditions. Still, quite often the data obtained lend themselves to descriptive, comparative, or correlational strategies.

ENGINEERING

The NGSS accord equal stature to engineering and science. Schools have a long way to go in order to establish a K–12 program in engineering, but the NGSS provide strong leverage for doing so. Schools in Britain have for many years offered courses in design technology, and for decades educators in the United States have attempted to emulate them.[3] At the elementary level, "design tech" challenges children to invent pull toys that wobble. At the secondary level, design tech might challenge a student to invent a light-sensitive switch so that an electric toy car will move and stop when signaled by a flashlight. The inclusion of engineering in the NGSS is perhaps the most promising crack in unity's armor. However, it is beyond the scope of this book to look carefully at the prospects of engineering in K–12 education.

K–12 *FRAMEWORK* FROM THE NATIONAL ACADEMY OF SCIENCES

The precursor document to the NGSS—*A Framework for K-12 Science Education: Practices, Crosscutting Concepts, and Core Ideas* by the National Research Council, provides an intellectual and scholarly foundation for the revision of the original National Science Education Standards (NSES). Its authors emphasize actual practices undertaken by experts in distinct contexts. Unfortunately, it appears that in translating this document into the Next Gen standards, the quest for unity co-opted its good insights. Fidelity to the diversity of actual practices among the sciences and the usefulness of core concepts in well-defined domains has gone missing in the three-dimensional superstructure of the NGSS.

The *Framework* rejects "the tendency to reduce scientific practice to a single set of procedures, such as identifying and controlling variables, classifying entities, and identifying sources of error."[4] According to the *Framework*, this tendency:

> Overemphasizes experimental investigation at the expense of other practices, such as modeling, critique, and communication. When generic procedures are taught in isolation from science content, they become the aims of instruction in and of themselves rather than a means of developing a deeper understanding of the concepts and purposes of science. Second, a focus on practices avoids the mistaken impression that there is one distinctive approach common to all science—a single "scientific method"—or that uncertainty is a universal attribute of science. In reality, practicing scientists employ a broad spectrum of methods.[5]

The quotation from the *Framework* quietly notes one of the most troubling aspects of the quest for unity: procedures taught in isolation become the aims of instruction themselves. In effect, the heuristic displaces content. The generic category scheme—whether labeled practices or processes—meant to illuminate different disciplines becomes the object of study in its own right. The components of scoring guides begin to look like valued content, as does the EQuIP and PEEC rubrics nationally and the Oregon Inquiry Scoring Guide locally. Many progressively minded science educators openly argue that inquiry is a content domain because there are many concepts and principles pertaining to inquiry to master. Keep in mind that the original NSES organized content in eight domains and that one of these domains was inquiry.

At least the NGSS have one anchor set in sound thinking about the diversity of methods deployed among the sciences and the hazards of reinforcing belief in the scientific method. However, even the admonitions presented in the *Framework* do not go far enough. The dominant imagery of scientific unity co-opts the *Framework's* attention to diversity, reducing its provocative message to no more than adding "modelling, critique, and communication" to the generic list of practices.

As the NGSS's precursor document, *A Framework for K-12 Science Education* stressed the importance of learning the actual practices of science together with disciplinary core ideas. As a call for revision of the 1996 NSES, the *Framework* presented a vision promoting depth of understanding over breadth of coverage and decried how little the superficial alignment of teaching with lists of standards accomplished to make science interesting. Disinterest and disenfranchisement, the authors argued, followed when students encountered facts in isolation and where they gained little knowledge related to their personal lives. The *Framework* stressed how actual practice pursued matters of social importance as well as how methods of inquiry responded thoughtfully to particular problems. There's quite a good level of respect for diversity in these principles. They are part and parcel of the NGSS's heritage and not to be overlooked.

DISCIPLINES AND VALUE

The NGSS crosscutting concepts deliberately scramble the disciplines and thus lose sight of the question, "What makes a discipline a discipline?" There are thousands of journal titles in thousands of fields. Perhaps there are as many disciplines as there are investigators engaged in truly pioneering work. Rampant disciplined study abounds.

On the other hand, there are precious few disciplines in school science. If not the common habits of mind, planning school science does require a principle that includes some topics while excluding others.

Disciplines begin as interests, and the spontaneous interests of children resemble these starting scientific points. As investigators pursue their different interests in depth, they invent productive ways of thinking and useful methods of inquiry. Disciplines emerge as those with shared interests coalesce around trusted ideas and share successful techniques of investigation. Experts proceed to teach novices how to discipline their inquiries in similar fashion.

Disciplines mature as their respective methods of inquiry prove fruitful in a wide array of contexts. They prove themselves responsive to a small number of enduring questions. How did life begin? What are the universal and fundamental constituents of matter? What is energy? Why do some organisms live socially? How did the earth get its crust? What rules govern fluid dynamics? The list is long, but not endless, and it is very interesting.

Still, schools cannot teach everything; the avalanche of disciplines needs disciplining. Schools need to teach, for example, disciplines in the service of achieving a sense of place and in the cause of attaching citizens to the husbandry of their local landscapes. Public schooling as a public good must encourage a sense of community membership, of identity through participation. An ethic of caring and a commitment to living responsibly within democratic communities ought to top the list of values shaping public education, even the science curriculum.

A vacuum exists once the unifying features of science have ceased to serve as the key to constraining and prioritizing "what to teach." Concern for "what to value" and "what holds interest" should guide filling this vacuum. Interest is basic. Understanding has value. Disciplines that demonstrate value to personal and community life, learning that cultivates interests and teaches how to pursue them in depth, should drive the curriculum.

The choice of disciplines to study and the range of interests to cultivate will vary by community and region—even from school to school and, in significant ways, from student to student once the top-down dogma of universal science has been laid to rest. De-emphasizing process in answer to "What to teach?" invites a bit of chaos into organizing school science. However difficult and messy it may be, answering "What to value?" offers direction and constraint. There is no escape from returning to questions of value when designing curricula expected to serve society well.

Babies and Monkeys

People value good medical care, and good medical care depends upon medical research both with humans and nonhuman primates. One of the nation's primate research centers sits in a suburban neighborhood in Beaverton, Oregon. Few local citizens are aware that thousands of monkeys live nearby and not at the zoo. From time to time, animal-rights protests capture headlines, and scientists at the primate center rightly worry about whether public sentiment will someday bring their research enterprises to a halt. The value of medical research as a means to alleviate human suffering and promote healthy living justifies in their minds, as well as among funding agencies of the US government, experimenting on other species—but not everyone else agrees.

Testing new forms of contraception on nonhuman primates goes well beyond the parameters of school science, but adolescents are perfectly capable of grasping the role of ethics in medical research. The prominence of ethics in this field's research protocols distinguishes the science from others. The topic has special significance to the community that houses the primate research center. It's an economic resource and a political lightning rod. Medical research with human and nonhuman primate subjects such as Rhesus macaques crystallizes value and ethical questions at a very personal level.

A recent multiyear project immersed middle-school teachers from social studies and science departments in the anthropological study of medical research for a short time.[6] Teachers returned to their classrooms anxious to introduce ethical and value questions pertaining to basic research with primate subjects and clinical human trials in medical science to their students. Each teacher prepared a unit of study to introduce their students to the medical research community's culture (or "cultures"—the practices of the primate lab and clinical studies did differ on many levels). One science teacher, for example, presented to his class a proposal to test a salt substitute on infants:

> He posed a problem that called for investigation with human subjects: "One of the most irritating parts of eating fast food is the pain caused by salt from the French fries entering small paper cuts on the hands. This study examines the amount of pain involved with pouring different types of natural salts into paper cuts on the hands of human babies. If successful, we hope to identify less painful alternatives to regular salt."

Keep in mind this was a hypothetical study. No one actually inflicted pain on human babies as part of this exercise. The teacher intentionally proposed an outrageous study as a thought experiment for getting students to express their thinking about ethics in the context of clinical trial, medical research.

He continued his description of the clinical trial's design, ending with a definition of the dependent variable:

> One thousand newborn human babies will be randomly selected from local hospitals. Each baby will be given a paper cut on the palm of his or her hand. Four groups of 250 babies each will then be given French fries with the following different types of salt used to flavor the fries:
>
> Group 1: Standard table salt (sodium chloride, NaCl)
> Group 2: Iodized table salt (sodium chloride with iodine)
> Group 3: Sea salt (various natural salts with assorted minerals)
> Group 4: Salt substitute (potassium chloride, KCl)

> If the babies cry when they get the salt on the paper cut, that will be a positive test for pain. If the babies don't cry, that will be considered a negative for pain. After the study is completed, babies will be sent back to the hospital.

The study certainly sought a robust sample size! There would be very many babies to wound. Perhaps "informed consent" needed some attention.

Babies certainly cannot give informed consent; the ethical dilemmas are just getting started. Of course, middle schoolers were outraged by the idea of inflicting pain on babies. More subtle was the ethical issue of giving babies salty French fries to eat.

Medical researchers do not begin studies of the efficacy of new treatments on large samples of humans, especially children, prior to a great deal of research of a different sort. Sometimes the problem justifies an animal model.

Would students feel any different if the research protocol were designed to test painless salt on monkeys? Monkey parents cannot give informed consent for their youngsters to participate in clinical research, either. Moreover, unless a study is conducted to answer questions of veterinary practice with monkeys, there is no benefit in return for the induced suffering. Who speaks for the monkeys?

The teacher's exercise worked well as a representation of the nature of medical science research—its distinctive obstacles, its value context, and its research protocols. It stimulated animated discussion, to say the least. He clothed the exercise in the language of classical single-manipulated-variable experimentation. He might have gone further to address the issue of human variability, even among infants, in relation to randomization in the design of medical research.

The middle-school lesson, "Bring Us Your Babies: Using Controversy to Teach the Value of Science," became popular among others who had worked with him in the project. It demonstrated that there is room for teaching distinctive features of disciplinary practice within the dominant paradigm of experimental science at the middle-school level.

Banishing Naiveté

"What to value?" inescapably frames discussion about "What to teach?" For many decades the "habits of mind common to all the sciences" and the "processes [now practices] characterizing inquiry" have dominated as the answers to "What to value?" Displacing them creates a vacuum where debate over what to teach may resume, even if very intense.

Honoring diversity opens the door to a variety of disciplinary concerns and each utilizes methods appropriate to the demands created by its interests. Excluding this diversity diminishes the potential to attract students into the sciences. The prospects for geological, meteorological, and archaeological sciences to contribute to K–12 education, for example, remain largely untapped.

Tangible methods of inquiry prove their worth in context. There, concepts themselves function as tools of inquiry—homology in biology, frame of reference in physics, for example. Properly taught, disciplinary core progressions lead to respected performance—a proper expectation. Performances demonstrate concepts and methods put jointly to work. They are not illustrations of what scientists in general and in the abstract do.

There are disciplines that speak to social value very loudly: medical science is one; climate science, another. Often the sciences that address complex systems with many interacting variables bear fruits that are of high social value. Stereotypical science cannot prepare people to appreciate them. The approaches taken by field science in the study of animal behavior suggest how to teach even children in the primary grades about complex phenomena.

The NGSS do not preclude advancing in these directions, but they do hobble efforts and cause educators to spin their wheels unproductively. Making social and civic value a primary criterion for deciding what to teach invites serious treatment of ideas from many fields. The historical style, field studies in natural settings, the *Framework* document, engineering's status, and the value dimensions of many sciences have already cracked unity's armor. They point promisingly in the direction of banishing naïve ideas about scientific unity.

NOTES

1. Mark Windschitl, Karen Dvornich, Amy F. Ryken, Margaret Tudor, and Gary Koehler, "A Comparative Model of Field Investigations: Aligning School Science Inquiry with the Practices of Contemporary Science," *School Science and Mathematics* 107 (2007): 382–90.

2. Robin Burnett, *The Pillbug Project: A Guide to Investigation* (Washington, DC: National Science Teachers Association,1992).

3. Susan Dunn, and Rob Larson, *Design Technology—Children's Engineering* (New York: Falmer Press, 1990).

4. National Research Council Committee on a Conceptual Framework for New K-12 Science Education Standards, *A Framework for K-12 Science Education: Practices, Crosscutting Concepts, and Core Ideas* (Washington, DC: The National Academies Press): 43.

5. Ibid., 44.

6. William Cameron, James Parker, Mary Samuels, and Susan Shugerman, "Responsible Research Lesson by Steve Lent," *Teachers Immersed in the Experience of Science (TIES)*, Science Education Partnership Award (Portland, OR: Oregon Health & Sciences University, 2006–2010): Grant Number R25 RR020443, National Center for Research Resources (NCRR)/ National Institutes of Health (NIH), http://www.ohsu.edu/xd/education/student-services/upload/TIES-ResponsibleResearchLesson.pdf (accessed February 13, 2015).

15

The Promise of Place

At Kilchis Meadow High School, perched above the Columbia River estuary, students farm crawfish. They also work on restoring local salmon runs and share their opinions on state law governing who can catch how many. The school serves a sparsely populated county where the economy of natural resource extraction (timber, fish, and grazing land for dairy cattle) dominates. All of these enterprises have been in decline, and unemployment in the county is high.

During the Corps of Discovery Bicentennial celebrating the Lewis and Clark expedition, the school's principal encouraged cooperation among the school's biology, history, and English teachers. They brought on board a musically talented naturalist who developed a curriculum that the students followed for several years. The naturalist, dressed in period garb as George Gibson, the Corps of Discovery's most versatile musician, teamed with a fiddle-playing friend to add a living museum dimension to the curriculum. The fiddler took on the persona of Private Pierre Cruzatte, the one-eyed, French-Omaha boatman on the expedition who unceremoniously happened to shoot Meriwether Lewis in the buttocks.

The Kilchis Meadow curriculum, supported by a grant from the Columbia Estuary Program, dug deeply into the history of the place the students inhabited. The land was an active plate margin dotted with ghost forests where lavas that erupted from fissures hundreds of miles to the east had come to rest in submarine canyons now pushed skyward.

On the north shore at the mouth of the Columbia rises Cape Disappointment, named in 1788 by Captain John Meares, who, being deceived by waves breaking over the bar, wrongly concluded that no great river reached the Pacific at this latitude. A few years later American efforts to enter the fur trade in the Far West led to the establishment of Astoria on the Oregon side of the river.

Bountiful runs of anadromous salmon (including infamous "June hogs"), steelhead, smelt, and lampreys migrate the river—a protein source that has supported human culture for thousands of years: the Cathlamets, the Clatsops, the Wahkiakums, and other Chinookan peoples, then European American settlers, fishers, and traders.

The Bicentennial curriculum achieved coherence through story, experience, and context. History and science alternated in exposing the strata of the past and forecasting the future. The curriculum put disciplines to work toward achieving a sense of place. Threading through the stories of the local landscapes and communities of the Lower Columbia coursed the wisdom of its first peoples, the surge of European American expansion, the ecological interpretation of Cascadia, and the geologic history of western North America.

PLACE AS STRATA OF STORIES

The journals of the expedition provided the curriculum's spine. Journaling, taught the English teacher, was a respectable genre of writing—and students learned to keep journals as they worked in the field to replicate experiences recorded by Lewis and Clark. Lewis's journal included sketches of the oily candlefish, eulachon (or "ooligan," an anadromous smelt), and vegetation.

The Kilchis Meadow students sketched what they found, which meant documenting the distribution of nonnative, invasive species such as purple loosestrife. Soon they were working with county agents to disperse *Galerucella* beetles as biological controls of loosestrife. Two species of these beetles eat leaves and stem tips. Root-eating weevils, *Hylobius transversovittatus*, are also used to control loosestrife.

Near the school on the Julia Butler Hansen Refuge roamed a herd of Columbian white-tailed deer, an endangered species. Prior to European immigration, the deer ranged widely in the bottomlands of the Willamette and Columbia watersheds, making them exceptionally easy to hunt. Today, compressed into living in a small area with extremely high rainfall that depletes the soil of vital nutrients, they subsist on a diet especially lacking in selenium. Fawn survival has been worrisomely low, and coyote predation is a problem.

As the tides ebb and flood the estuary's marshes that ring the refuge, juvenile salmon fatten themselves on small crustaceans before completing their journey to the Pacific. They linger in the estuary as their bodies acclimate to saline conditions.

Humans recently scavenged these marshes for nonnative clams—in the same waters where Native Americans had harvested wapato—until health authorities declared them unsafe to consume in large quantities. Voracious Caspian terns, a protected species, decimated the numbers of young salmon, another protected species, until wildlife managers managed to force their relocation closer to the mouth of the river where the terns could vary their diet. Now populations of cormorants have taken their turn to feast on juvenile salmon, creating a new dilemma for wildlife managers.

Channel Navigation

The US Army Corps of Engineers dredges to keep the channel open, deepening it for Panamax-class ships—container and grain haulers with the maximum draft for passage through the Panama Canal. Rows of new homes line the Washington shores of the Lower Columbia, unprotected by the levees behind them. Recreational living has become a new growth industry. To replenish their beaches, these new homeowners avidly seek dredge spoils, sediment that in Lewis and Clark's day was flushed to the ocean by winter runoff. Engineering has become a key component of the local ecology.

Lewis and Clark mapped the area as they paddled through. The students traced the water route in double kayaks and learned to map it by taking bearings carefully and estimating distances. They found once-navigable channels now choked with sediment, the consequence of clear-cutting along the river's tributaries. Lewis and Clark also encountered a landscape of villages emptied by smallpox. Malaria added to this toll decades later.

A Chinook leader, Chief Comcomly, received a medal from Lewis and Clark in recognition of his skill as a bar pilot on the Lower Columbia. Sediment bars are not novel, just more common farther upstream than in Comcomly's time. When Chief Comcomly died, his skull, flattened by being bound to a cradleboard while he was an infant, in the fashion of upper-class Chinooks, was shipped to England. Phrenologists of the era gave great credence to skull morphology.

As told by Russell Shorto in *Descartes' Bones*, bumps indicated the relative development of different brain areas devoted to "guile, courage, tendency to commit murder, sense of proportion, architectural talent, sense of satire, benevolence, obstinance, and language ability."[1] Facial angles, skull bumps, nose bridges: various "scientific" research programs examined them all in hopes of finding physical evidence of mental and character traits associated with a person's face in profile as well as the stereotypical traits of an ethnic group.

In 1830 Chief Comcomly's skull and skills were a scientific anomaly in need of an explanation. His forehead angled sharply to the back, the exact opposite form that Europeans associated with intelligence. Europeans venerated a Greek and Roman straight profile.[2]

Scientific racism has been no stranger to the Lower Columbia. Chinese laborers suffered from discrimination in the late nineteenth century and were prohibited from entering the United States by the Chinese Exclusion Act of 1882 until its repeal in 1943. Chief Comcomly's story opens a window into the history of scientific racism in the region. Following a peripatetic museum life, Comcomly's skull was returned to his relatives for burial in Ilwaco, Washington, in 1972.

Salmon, Oyster, and Stickleback

Once the fur trade had faded and disease had decimated Chinook populations, cutting-edge technology created a new economic boom along the Columbia: canning, especially of the June hogs, also known as Chinook salmon. As the cannery production feeding the eastern population of laborers peaked, fish wheels and gill-netters vied for rights to the catch. Each group argued for the moral superiority of their way of life. Gill-netters experimented with seining the river with nets pulled by horses at low tide. Most ventured out in small boats remembered for their sails as the "butterfly fleet." Catches dwindled. Restoration projects now keep wild salmon from extinction. Their effects mingle with changes in food abundance in the North Pacific. Kilchis Meadow students learn how the salmon link terrestrial biome and ocean ecologies.

Finns, Swedes, Icelanders, and the Norse migrated to the Lower Columbia to fish and fell timber; Chinese workers labored in the canneries. Many—often in their teens—perished in their labors; mortality and morbidity records housed in the county courthouse document how. Canneries and clear-cuts were dangerous places to work.

From the perspective of place, experience gains in coherence. For the student, learning about the local economy and history engages with an understanding of ecological, geological, and meteorological phenomena. The landscape brings these elements together. Coastal communities dominate Washington's Wahkiakum and Pacific counties. Tides and mudflats, shellfish and migratory birds, fisheries and canneries, and rain—lots of rain—coexist. The coherence stemming from inhabiting a particular landscape differs from the coherence found in the structure of a discipline.

These two types of coherence may reinforce each other, and one should not displace the other. When they mingle, learning science must share the limelight with other topics, including music and history. Disciplinary coherence—subject knowledge—need not trump experiential coherence—sense of place.

Oyster farming took hold in Willapa Bay, and during the Second World War, women ran the oyster business. They worked in the mudflats at low tide regardless of the hour. At night, when air raid sirens sounded, they had to extinguish their lanterns. A mile from home in blackness on black mud with the tide rising was very dangerous. On shore, women honked car horns to guide the oyster workers to safety. A local folk band has commemorated this era with the lyric, "The neighbors would honk us all home in the rain."[3]

Mudflats provide bountiful diets for shorebirds. *Spartina* marsh grass, imported from the East Coast as packing material for shipping oysters, escaped from warehouses and expanded aggressively, placing the oyster business as well as shorebird foraging in jeopardy. Local oyster farmers now work diligently to control its spread.

Field study in the marsh starts with beetle and weevil release to control purple loosestrife and ends with insight into the relationship between evolution and development. Here a small fish, the three-spined stickleback, scoots through the marsh shallows. A single gene's inaction in this fish can suppress the growth of pelvic spines.[4] That's a good thing in muddy-bottomed, freshwater habitats where dragonfly larvae prey upon wandering sticklebacks. Smooth backs give predatory dragonfly larvae a little less to grab onto. In open water, the gene turns on to promote the development of pelvic spines. Pelvic spines deter predator fish, who learn to avoid hooking their mouths on stickleback barbs.

Sentinels and Parapets

Stands of ancient Sitka spruce, stunted by growth in oxygen, and nutrient-poor soils, line the sloughs of the Lower Columbia. The lumber industry passed them by, and their spires seem oddly dispersed. Abundant shrubbery separates them; there is no closed canopy, but the trees are old. An occasional massive sentinel does rise above the others. Such sentinels stood here in January 1700, when the last Cascadia Subduction Zone great earthquake occurred. Land downstream near Skamokawa suddenly dropped by a meter, and water inundated the roots of Sitka spruce along the estuary's sloughs. As sediments accumulated, some spruce trees were able to grow new sets of roots close to the new surface. The shifting channel had toppled one such giant. It lay prone; its double root wads, a meter apart, were exposed at its base.

Mosses thrill to the pervasive wetness of the Pacific Northwest. Mosses grow on rocks, and mosses grow on logs. They drape the arms of the forest sentinels. Moss trap carbon dioxide and increase its concentration close to the earth. Mosses, like a rainforest, develop vertical structure with habitats arrayed from moist below to arid above, their sporophyte generations flying kite-like in the wind, the better to disperse spores. Crane fly larvae, the moss forest's sloths, ply the folds of vegetation. Fierce but tiny water bears—the micro-leopards of the moss rainforest—stalk nematode and rotifer prey. Fundamentals of ecological theory bed down on mosses.

Bedrock cliffs of frozen lava, festooned with mosses and wildflowers, tower over a portion of the river, parapets of flood basalts that traversed hundreds of miles to reach the Pacific fifteen million years ago. They are the frozen offspring of a mantle plume hotspot now named Yellowstone National Park. Footprints of the plume that scorched Central Oregon long before Yellowstone country existed go by the names of Crooked River and Wildcat Mountain calderas.[5]

These lavas once filled a submarine canyon. Now the canyon stands in high relief, and thanks to the mechanics of an active plate margin, the softer, fossiliferous rocks around it have eroded. The skyline echoes how North America and the Juan de Fuca plates in the Pacific Ocean have bulldozed against each other for tens of millions of years.

PLACE AS CROSSROADS

The Corps of Discovery Bicentennial curriculum drew upon the rich strata of stories typical of a particular place. Vibrant communities have come and gone, shifting their resource base and adapting new technologies, a thread that runs unbroken from present to past. The town of Skamokawa ("Smoke on the Water") is just now shaking off its recent, dilapidated history. It was once a Venice-like community made prosperous by the cannery, dairy, and timber industries. Skamokawa was roadless for much of its history but well equipped with rowboats, the favored means of commuting to school.

Other towns of the Lower Columbia have vanished, both Chinook and European. Occasionally blue trade beads from two hundred years ago turn up on the abandoned beach, referred to locally as "the beach of heaven." At one Cathlamet village site, fourteen cedar lodges housed several hundred people. At the ruins of a cannery ghost-town site, a local resident found a decorative blue bottle. The small bottle was once filled with the tears of a mourning Victorian-era mother, the wife of a cannery "baron."

Teaching through Place

These stories, taken to heart by Kilchis Meadow High School students, mesmerized a group of visiting teachers from the city in a weekend seminar. "Until I learned these stories, I thought Skamokawa was just another pleasant place to bike through and maybe stop for a sandwich," remarked one of them. Through the study of place, Kilchis Meadow students extracted meaning and built identity. Archaeological digs and settlement history gave context to the invasion of nonnative plants and animals, the decline in salmon runs, the legacy of logging, the scale of the canning industry, and the importance of dairy farming. High-school students enthusiastically shared these stories of place with the visiting teachers.

Teachers and students also gave their attention to the problems of the day: the geologic hazards of tsunamis, earthquakes, and landslides; deepening the shipping channel; balancing the needs of competing protected species of wildlife; and promoting new forms of commerce including tourism, which is improving, and hybrid poplar farming, which is not doing well. Some in their economically distressed community looked with favor on a proposal to construct a liquefied natural gas export facility; others objected strongly.

Teachers from the city learned how the biology, history, and English teachers from a rural community had assembled a coherent curriculum by mining the strata stories for interpreting the local environment. Its historic school housed a museum and small gift shop where local folk musicians peddled CDs of songs about oyster harvests and abandoned fishing vessels. Skamokawa, they realized, was a crossroads both scientifically and historically. They found estuarine ecology, plate margin tectonics, and invasive species management juxtaposed with stories of lives lived generation after generation in dramatically different fashions—those living in the present, like those who lived in the past, trying to figure out how best to fit the landscape. The depth of the stories gave pause . . . and hope.

Place as Counterpoint

Such is the promise of place-based education as counterpoint to standards-based education. The two are not incommensurable, but the latter has a grip on policy.

The coherent experience of one's own backyard, enriched when viewed through multiple disciplinary lenses, needs no boost from crosscutting concepts. Moreover, the methods of investigation are first and foremost the ones that work in the local context—disciplines in action, not abstraction.[6] In fact, place-based educators must avoid reducing the sciences to generic processes. Giving priority to social issues may tempt them to do so.

In the place-based curriculum, the particulars of local story illuminate wider history and underscore the importance of ecological and geological knowledge. In all places threads intersect, tying local events to bigger pictures. All places lay down strata of stories, from the bedrock of the earth to the current underpinnings of the economy. Each generation composes the landscape with fresh insight. The interpretation of place calls upon scientific and other disciplines, putting them to service in the lives people live and offering a promising solution to the limitations of the standards movement.

NOTES

1. Russell Shorto, *Descartes' Bones: A Skeletal History of the Conflict between Faith and Reason* (New York: Doubleday, 2008).

2. In 1811 Chief Comcomly's daughter, Ilchee ("Moon Woman"), married Duncan McDougal, chief factor of Jacob Astor's Pacific Fur Company. A bronze statue of Ilchee along the City of Vancouver, Washington's Waterfront Renaissance Trail honors the Chinook people, as does the restoration of the Cathlapotle Plankhouse at the Ridgefield National Wildlife Refuge. The Chinook continue to conduct tribal business and ceremonies at the plankhouse.

3. Mary Garvey, "Oystershell Road," recorded by Willapa Hills, *A Portrait of Place: Songs of the Columbia-Pacific Region* (Inn at Lucky Mud, 2010).

4. Sean B. Carroll, *The Making of the Fittest* (New York: W. W. Norton & Co., 2006).

5. Jason D. McClaughry, Mark L. Ferns, Martin J. Streck, Karyn A. Patridge, and Caroline L. Gordon, "Paleogene Calderas of Central and Eastern Oregon: Eruptive Sources of Widespread Tuffs in the John Day and Clarno Formations," in Jim E. O'Connor, Rebecca J. Dorsey, and Ian P. Madin, eds., *Volcanoes to Vineyards: Geologic Field Trips through the Dynamic Landscape of the Pacific Northwest: Geological Society of America Field Guide 15* (Boulder, CO: Geological Society of America, 2009): 423–50, doi: 10.1130/2009.fl d015(20).

6. Charles R. Ault, Jr., "Achieving Querencia: Integrating a Sense of Place with Disciplined Thinking, *Curriculum Inquiry* 38 (2008): 605–37.

Conclusion: A New Coherence

The unifying dimensions of the Next Generation Science Standards (NGSS) promise coherence and direction for the core disciplinary ideas of K–12 science education. At the same time the standards disassemble disciplines and then rearrange them as a matrix suitable for gauging the progress of all learners toward common ends. The approach resembles that taken by a large corporation prioritizing goals and then expecting all of its managers and employees to work as a team toward their realization. The approach assumes that fierce competitors (other nations) are doing the same. The result is "Science Ed, Inc.," an all-American brand. In combination with state governments, the corporate style of management-by-objective enhances the bureaucratic regimentation of school learning, a long-standing problem.

That's a bad way for a democratic society to bet on the future, to cultivate diversity, respect liberty, encourage creativity, and promote resiliency. Bureaucratic regimentation and its corporate-style marketing run roughshod over personal interest and eviscerate bodies of knowledge in a rush to inculcate all Americans with the same scientific dogma. Society ought to value diverse outcomes from learning the sciences and reject the generic imagery of a universal enterprise. Not only is the price of K–12 coherence and direction as defined by the NGSS too high, but the primary categories setting direction are mistaken.

Unfortunately, they misrepresent the match between thinking and doing in real-world contexts, straining to shoehorn all sciences into the same abstract categories of scientific practices and crosscutting concepts. Whatever the debate within the history and philosophy of science regarding scientific inquiry, the unifying constructs carry pedagogical liabilities for K–12 education.

Coherence is essential to meaningful learning. There are two sources of coherence more appropriate to K–12 curriculum design than scientific universals: disciplinary structure and personal experience. The teacher's task is to act as a bridge between the two, as Dewey clearly explained over one hundred years ago.

Disciplines—trusted heritages of knowledge and methodology—offer direction for the growth of experience. Disciplines are not inert; they branch, coalesce, and simultaneously discover more about the world while learning more about how to investigate the world.[1]

Several approaches stand ready to emerge from the shadow of the quest for unity and assume the mantle of coherence. Each in its own way embodies a blend of disciplinary and experiential approaches to curriculum design.

"A-DISCIPLINARY" THOUGHT

Bridging between the established sciences and personal experiences may begin with "a-disciplinary" thought, that is, engagement in problem-solving and inquiry without organization imposed by disciplinary structures. Topics that organize the experience of science during childhood ought to be instrumental to the achievement of elementary educational aims, not the means to master abstract schemes of scientific unification. Nothing is more fundamental than cultivating and disciplining interest. Gaining confidence in oneself as a learner is indispensable. Developing aesthetic appreciation for the natural world matters greatly. Literacy and mathematical proficiency count, too.

Disciplining thinking begins with asking whether the method of inquiry is adequate to solving the problem at hand—something a child can do when confronted with the problem of how many snails live in the fish tank or where the black squiggly stuff on the bottom came from.

Good curriculum, writes Deborah Seltzer-Kelly, asks children "to engage the natural world in all of its variety and innovation."[2] Interest is basic, whether spontaneous or carefully prompted by a teacher's actions. The *Elementary Science Study* exploited the "interest is basic" principle very effectively. It stressed exploration as a context for generating questions and provided teachers with insights to guide children's more deliberate investigations.

Children engaged in ESS *Optics*, for example, projected colored light, giving them the opportunity to form colored shadows and blend, bend, and mix beams of different colors. Their explorations focused attention on the challenge, "How can I make white light?" Today science museums often feature exhibits where visitors can mess about with light and make shadows of different colors. They find the experience intrinsically motivating.[3]

ESS designers kept the phenomena of interest center stage. Immersed in inquiry activity that exercised language and imagination, children in ESS classrooms gained satisfying insights into balance, simple electrical circuits, colored solutions, pendulum motion, spinning tables, and much more. ESS modules made use of common organisms: mosquito wrigglers, earthworms, and mealworms.

Skilled teachers, using ESS-style materials, continue to value exploratory and open-ended inquiry as a context for forming new concepts—ideas to use and, through use, gain insight into their meaning. They arrange ample time for children to talk science. Teachers who fail to use inquiry as the context for developing ideas reinforce the reduction of science to process.

KEEPING IT REAL

Keeping close to real-world examples helps to promote interest while honoring what disciplines do best. For example, it's better to focus attention on replicas of actual footprints of extinct beasts than to re-craft them as mysterious footprints in the snow. Real paleontologists look for analogues of dinosaur behavior in the footprints and antics of living beasts.

"Common themes" constitute a "coherent vision of the knowledge and skills [for] every high school graduate," claim the standards enthusiasts.[4] Making "common themes" the vision of coherence for all treats disciplinary particulars as afterthoughts or interchangeable parts. The common themes eclipse how the tools of the paleontological trade, or any other field, prove themselves worthy.

The footprints puzzle began as an exercise for teaching about dinosaurs in the 1960s. It has been used every decade since for a host of other purposes—all tied to teaching scientific universals, even at the zoo. Teaching for an understanding of universals undervalues discipline-embedded questions such as, "How do paleontologists figure out the behavior of dinosaurs from their footprint fossils?" There are answers that satisfy curiosity just fine. There is no need to justify them as illustrative of the themes and habits of mind spanning all of the sciences.

Analogies between the dead and the living exhume the secrets entombed in fossils. Modern rhea bones and rhea motion reveal how the Jurassic's *Allosaurus* roamed. African muds record the dancing steps of fighting hippos, the stealthy moves of stalking predators, and the panicked strides of frightened prey. The sanitized diagram of footprints in school science disposes of all this information. It's not the only exercise that does so.

A FOREST PARABLE FOR EDUCATORS

In a cautionary tale about the history of German forestry, James C. Scott explains the detrimental effects of stripping forests of their wild, organic complexity in order to manage them for efficient, commercial production. Initially foresters "hunted" for the right wood for ship masts, furniture, houses, and wagon wheels. Surveying, marking, and mapping the locations of trees suitable for different uses facilitated the hunt. Once the forest was completely cut, planting a new one began. The new forest would be better: a more logically organized, legibly arranged, planned forest:[5] a forest on a grid. The demands of "legibility," Scott concludes, drive forest design.

Mapping was now the first step, not the last. Foresters planned straight rows of single-species, same-sized, same-aged trees that made harvesting more efficient and the most valued woods more abundant. After successive harvests, however, the soil gave out and the grid-made forest failed. Standardization killed the forest.

Nowadays foresters practice more ecologically intelligent forestry. They value species diversity, soil microbes, and rodent vectors. They leave standing snags as perches for predators and piles of leafy branches to rot and recycle nutrients. Scott's forestry story is not a preamble to an attack on unsustainable forestry. Instead, it is a parable about the harm brought about by overzealous haste to standardize and measure. The harm done by the modern state in rational pursuit of its ends is Scott's concern. Educators should take note.

States seek to refashion the wild thickets of learning with grids and metrics that enable efficient management. They therefore impose a finite set of variables in the abstract, like the virtual forest, in substitution for the endless variety of real-world learners and chaotic array of things to learn. Management takes aim at achieving grade-level proficiency for all, and this aim sets in motion a cascade of consequences. Workers of the educational research world unite to measure the effects of instructional treatments on learner proficiencies. They favor "legible" learners, legible treatments, and legible proficiencies—and turn to the sciences for assistance in bonding causes to effects.

The geometry of a tree plantation pales in comparison to the structural sophistication of standards documents. They dissect and cross-reference, sequence and subsume, align and categorize knowledge in presumably logical and legible hierarchies. As a result, teachers and students find themselves trapped in a seemingly endless maze of alignment protocols.

The standards document, in the role of the planned plantation map that replaces the wild forest, recasts the complex landscape of learning as a three-dimensional grid. The great irony is that the unifying dimensions of the NGSS defy measurement *and* sterilize the fertile soil the mind depends upon for growth.

ACHILLES' HEEL?

Those who reject the *Challenging Science Standards* argument critical of the quest for unity may believe it has an Achilles' heel: claiming universality for its own abstract, principles of science. The argument does repeatedly emphasize the need for matching methods to conception in order to meet the challenges of inquiry. Is the argument against unity self-contradictory?

Premising science teaching on the "challenge of the match" aims to counter teaching the sciences as unified by habits of mind and processes of inquiry. There are three counterpoints to the charge of self-contradiction in the counterargument. First, the "challenge of the match" principle functions as a starting point, not a conception of finished understanding. It is a first step in the search for diversity. The two unity dimensions of the NGSS, on the other hand, lay out both the starting and end points for science instruction. In addition, they bend the distinct differences among core disciplinary ideas toward these points.

Second, the "challenge" principle is a rule of thumb—one to be trusted, certainly, and an insight worth adopting. In contrast, scientific practices, process skills, understandings of the nature of science (NOS), and crosscutting concepts in the standards are elaborate heuristics guiding what to learn from and about science. They become, in effect, a content domain in their own right.

Teachers have long approached the scientific method from the perspective of a heuristic turned into content. "I really am most interested in having my students learn the scientific method," is an age-old refrain. The original science standards created several separate content domains out of inquiry heuristics. This proliferation changed the refrain to "What I care most about is that my students learn skills of inquiry" or "that my students understand the nature of science." In briefest terms, *Challenging Science Standards* warns educators not to substitute the heuristic for the content. Look, instead, for how disciplines discipline their inquiries on their own terms.

Third, the "challenge of the match" statement differentiates into: "match *appropriate* methods to *distinctive* conceptions in order to meet the *diverse* challenges of inquiry in *different* contexts." In effect, the adjectives added reinforce the principle to seek and appreciate diversity. Distinctively diverse conceptions and their corresponding methods do the work of different enterprises. They differ because problematical phenomena of interest in the world vary incredibly. What to do about the situation? Honor the diversity of disciplined thinking.

Disciplined thinking depends upon a small population of concepts for organizing thinking and doing in a purposeful context. Investigators invent and then test the worth of concepts as adaptations to solving problems. A set of successful concepts earns the label *conception*. Worthy conceptions endure, and on the conception all else depends. This focus has potential to animate the NGSS's progressions of core disciplinary ideas. Ideas need to progress somewhere interesting and useful, not become interchangeable parts for inculcating the dogma of unity.

DOGMA'S QUEST

Debate over universal criteria of good science and the status of different disciplines has existed for a long time. The quest for unity has proven popular since the inception of science education whether serving British gentlemen or all Americans. The quest has become a dogma able to seduce several constituencies. It gives the discipline of science educators an identity. It draws support from scientists arguing the scientific equality of their fields. And it suits the bureaucrat's appetite for standardization compelled by the drumbeat of accountability.

Values and Aims

In order to constrain the claims of the host of disciplines ready to storm the gates of school science, value judgments must be made. Preparing citizens for lives of social responsibility in a democratically governed society ranks at the top of the value hierarchy. Tied for that position is another value, one that has energized the standards movement: promoting equal opportunity in a competitive world burdened by embedded privilege.

Diversifying what schools teach promotes personal interest and responds to social value. Among young children, spontaneous interests and topic work—often with art, mathematics, and literacy being indistinguishable from science lessons—inevitably lead in many directions. As children mature, they deserve opportunities to acquire diverse disciplinary understandings in soil science, health and medicine, geomorphology, deep time, local ecosystems, and perhaps even taphonomy—not just in physical, life, and earth and space science subsumed under crosscutting concepts.

Science teaching needs to place the disciplines in service to higher aims. Among those deserving high status are:

- Finding pleasure in knowing the natural world
- Achieving a sense of place
- Gaining confidence in oneself as a clever learner
- Honoring the talents of others
- Valuing attachment to community
- Respecting the diversity of scientific enterprises
- Becoming committed to improving society
- Appreciating objects of scientific interest through aesthetic lenses

These aims ought to hold value equal to the long-standing ones of promoting equal opportunity, teaching scientific habits of mind, and filling the science and engineering pipeline. There needs to be more conversation about how disciplines differ and what resources they might contribute to the expanded set of aims.

Limitations and Flaws

The seven NGSS crosscutting concepts and eight practices (plus the NOS understandings) presumed common to the sciences cap a century or more of struggle to translate the work of many scientific enterprises into a timeless, placeless, and subject-less portrait. The effort lies crushed beneath the weight of its complex-coded, abstract superstructure. It is anything but legible. Parents and citizens cannot comprehend its nuances and derivation. Volumes of jargon and research reduce to "the" scientific method and its marketing.

Traditional content—Schwab's "rhetoric of conclusions"—lurks about, debasing debate over "What to teach?" State legislatures argue over inert phrases in declarative statements, more worried about what students might believe than whether they grasp when and why to trust or distrust different sorts of disciplinary claims.

Disciplinary diversity illustrates what counts as good science and subverts the stereotype of the scientific method and its tendency to deify the adjective "scientific." The sciences are certainly not context-free, neither in their cognitive agendas nor value dimensions. In the end, the quest for unity serves to advance the social status of scientists by inculcating masses of students in a particular scientific worldview—dogma in an elaborate disguise. This effort has failed.

The quest for unity has impoverished the public's ability to differentiate criteria for judging the worth of different sciences. Climate science, medical science, and ecological science, to name a few, adapt intelligently to their challenges. It is difficult, but important, to understand that different fields wrestle with causation and uncertainty in different ways because they must. One size does not fit all.

This misunderstanding extends to the study of schooling itself. There is danger in relying upon experimental models from agriculture, physics, and medicine to measure school improvement—on invoking the old bugaboo of stereotypical scientific method in the realm of high-stakes testing. Making education amenable to experimental study risks making educational events resemble the phenomena familiar to these other disciplines—the contexts where their models work best.[6] Children ain't hybrid corn or elementary particles. And poor performance is not a disease.

Federal funding compels many school reformers to embrace experimental methods in order to measure progress. They, and the public receptive to their message, have overly internalized the imagery of a singular science inculcated by K–12 schooling.

Those committed to gauging student progress in understanding the practices and concepts common to all the sciences face intractable obstacles of assessment. On the surface, the short lists of terms such as "cause and effect, modeling, change and stability, communication" offer direction. Dig deeper and there's very little tangible substance to draw upon. As a consequence, measures of performance revert to the default mode: testing "right answer" science, the core disciplinary ideas in their inert, propositional form.

The NGSS encourage educators to set "performance expectations" for students by framing core ideas with practices and crosscutting concepts. The broad abstractions that comprise these two dimensions of the standards are not up to the task. Instead, educators must look to the heritage of resources matched to the demands of inquiry disciplined by context.

Dogma and Trust

The NGSS unifiers skew curriculum, demanding that all courses of study lead to the same abstract ends. They compromise planning. Teachers, assessment experts, and curriculum designers must spend uncountable hours demonstrating alignment with these ends. Time spent aligning detours them from spending time organizing instruction to achieve diverse, high-level understandings within distinct fields of study.

Practices do indeed, as the NGSS claim, combine skill and knowledge. True, disciplines obviously share practices demonstrated to work effectively in multiple contexts. As contexts vary to encompass distinct phenomena, practices differentiate. A wide array of inquiry skills serve as the tools of the scientific trades. Disciplinary knowledge is the expertise to know when, why, and how to use which ones. Appreciation of this expertise is the basis of knowing when to trust and when to distrust scientists' claims.

A people poorly taught the diversity of what scientists do and why will succumb to belief in an oversimplified, scientific method that seldom works as advertised. Such misunderstanding is prone to misplace trust. Each round of reforming science education has redrawn the generic portrait of science in an effort to remedy the public's misunderstandings of the sciences. Yet the newly imposed obsessions are easily forgotten, readily replaced in the public mind by dim recollection of "the" scientific method. It's been that way for many decades. The time has come for educators to challenge dogma's quest for a universal portrait of the sciences.

NOTES

1. Philip Kitcher, *The Advancement of Science* (New York: Oxford University Press, 1993): 203.
2. Deborah Seltzer-Kelly, "Deweyan Darwinism for the Twenty-first Century: Toward an Educational Method for Critical Democratic Engagement in the Era of the Institute of Education Sciences," *Educational Theory* 58 (2008): 304.
3. Deborah L. Perry, *What Makes Learning Fun? Principles for the Design of Intrinsically Motivating Museum Exhibits* (Lanham, MD: Alta Mira Press/Rowman & Littlefield, 2012). The cover illustration features making colored shadows.
4. Jo Ellen Roseman, "Science for All Americans: Timely or Timeless?" *NSTA Reports* 20, no. 3 (2009): 3.
5. James C. Scott, *Seeing Like a State* (New Haven, CT: Yale University Press, 1998).
6. John L. Rudolph, "Why Understanding Science Matters: The IES Research Guidelines as a Case in Point," *Educational Researcher* 43, no. 3 (2014): 15–18.

About the Author

Charles R. Ault, Jr. ("Kip"), Professor Emeritus, for twenty years coordinated the science education components of the Master of Arts in Teaching program at the Lewis & Clark Graduate School of Education and Counseling in Portland, Oregon. His courses served K–12 educators and featured field study in geology and natural history. Ault studied science education and paleontology as a graduate student at Cornell University, which led to his interest in teaching about geologic time. Before responding to the call of academia, Ault enjoyed the privilege of teaching elementary- and middle-school children in Connecticut and Colorado. His recent scholarship has focused on the importance of a sense of place, the nature of geological reasoning, and the limitations of teaching science as process. Ault is currently working on his next book, *Do Elephants Have Knees?,* which introduces serious Darwinian ideas with whimsical children's literature. In addition to science writing, Ault has focused his retirement on learning to roll his kayak reliably.

www.ingramcontent.com/pod-product-compliance
Ingram Content Group UK Ltd.
Pitfield, Milton Keynes, MK11 3LW, UK
UKHW041630031026
15587UKWH00001B/177

9 781475 818475